Turning Plants Into Medicines

NIPA® GENX ELECTRONIC RESOURCES & SOLUTIONS P. LTD.
New Delhi-110 034

Turning Plants Into Medicines

Novel Approaches

T. Parimelazhagan
Formerly, Scientist in DRDO-MoD (GoI) India
Presently Associate Professor, Department of Botany
DRDO-BU Centre for Life Sciences
Bharathiar University, Coimbatore–641 046
Tamil Nadu

NIPA® GENX ELECTRONIC RESOURCES & SOLUTIONS P. LTD.
New Delhi-110 034

NIPA® GENX ELECTRONIC
RESOURCES & SOLUTIONS P. LTD.

101,103, Vikas Surya Plaza, CU Block
L.S.C. Market, Pitam Pura, New Delhi-110 034
Ph : +91 11 27341616, 27341717, 27341718
E-mail: newindiapublishingagency@gmail.com
web: www.nipabooks.com

For customer assistance, please contact
Phone: + 91-11-27 34 17 17
Fax: + 91-11- 27 34 16 16
E-Mail: feedbacks@nipabooks.com

ISBN: 978-93-58870-51-0

Composed and Designed by NIPA®.

DRDO-BU CENTER FOR LIFE SCIENCES
(A joint venture of Defence Research and Development Organisation, Ministry of Defence and Bharathiar University)
BHARATHIAR UNIVERSITY CAMPUS
COIMBATORE - 641 046
Phone : 091-422-2428154, Fax : 091-422-2425459
Email : director.drdobu@gmail.com

Dr. P. V. L. Rao
Scientist 'G'
Director
DRDO-BU Centre for Life Sciences
Bharathiar University
Coimbatore–641 046, Tamil Nadu

Foreword

The increasing global demand for herbal medicine and products warrants accelerated cultivation, marketing and conservation of medicinal plants. Traditional medicine is widespread throughout the world and is an integral part of each and individual ethnic culture. Increased use of medicinal plants will most likely be part of this evolution.

It is my great pleasure that Dr. T Parimelazhagan has done a great job in gathering considerable scientific information on herbal medicine from academic experts and research institutes. The articles in this book will definitely allow the readers to go through their research fields more deeply and motivate the young researchers to do something novel for the society.

The book has given a good combination of traditional medicine and its current relevance with scientific approaches.

I am sure that the book will provide a platform to students and researchers to gather knowledge so that a multidisciplinary research program can be formulated for improving the research in phytomedicine.

P.V.L. Rao

Preface

Herbal medicines are thought to be the only remedy for the disease treatment during ancient days before the introduction of modern medicines. They formed the only accessible and affordable source of medicines in those days. Nature always stands as a golden mark by providing natural products from plants. The book '**Turning Plants Into Medicines: Novel Approaches**' has been designed to collate the prominent latest information and review the viewpoints of eminent investigators who excel in the areas of herbal drug research. This book is a compilation of 31 articles by eminent academicians and scientists on different aspect of the subject. All the recent scientific innovations/advances have been incorporated in a simple and lucid manner. A special emphasis has been laid on screening of herbal drugs for pharmacological activity, antioxidant potential, Phytochemistry, ethnobotany, ethnopharmacology, phytochemistry, bioinformatics and biotechnology involved in herbal research. The review processes of the articles have been carried out by experts from Universities and Research Institutes. We hope the present compilation will be useful for the Students, Research Scholars, Academicians and Industrialists and people associated with herbal research. We appreciate the efforts of all those who contributed the quality papers by sharing their current scientific knowledge. The editor would like to convey sincere thanks to Dr. G. James Pitchai, Vice-Chancellor, Bharathiar University for his guidance and encouragement. The support and guidance rendered by Dr. S. Manian, Professor & Head, Department of Botany, Bharathiar University is also acknowledged. The author would also like to thank his research team Dr. R. SenthilKumar, Blassan P. George, M. Iniyavan, K. Arunachalam, S. Saravanan, Rahul Chandran and Sajeesh T. for their contribution in the compilation of the manuscripts in a well designed manner. We extend sincere thanks to Department of Science and Technology, Ministry of Earth Science, Defence Research Development Organisation (Govt. of India), Tamil Nadu State Council for Science and Technology (Govt. of Tamil Nadu) for providing financial support to conduct "National Conference on Phytomedicine". Constant support provided by our colleagues of our Department of Botany, Bharathiar University, Coimbatore is also acknowledged. We would like to express our special appreciation for the publishers and their team for the sincere efforts in bringing out the book in time.

Editor

Contents

List of Contributors

A. Rajendran
Department of Botany
Centre of Life Sciences, Bharathiar University
Coimbatore – 641 046, Tamil Nadu

A. Babu
Centre for Plant Biotechnology
Department of Plant Biology & Plant Biotechnology
St. Xavier's College (Autonomous)
Palayamkottai, Tamil Nadu

A.Gangaprasad
Department of Botany
University of Kerala, Kariyavattom
Thiruvananathapuram, Kerala

A. Mahalakshmipriya
Department of Biotechnology,
Dr. N.G.P Arts and Science College
Coimbatore – 641 048 , Tamil Nadu

A.Siva Raman
Centre for Plant Biotechnology
Department of Plant Biology & Plant Biotechnology
St. Xavier's College (Autonomous) Palayamkottai
Tamil Nadu

Arun K Das
Department of Plant Science
Madurai Kamaraj University
Madurai – 625021, Tamil Nadu

Aruna Sharmili
Department of Biotechnology
University of Madras, Guindy Campus
Chennai – 600 025
Tamil Nadu

B. Meena
Department of Medicinal and Aromatic Crops
Tamil Nadu Agricultural University
Coimbatore – 641 003
Tamil Nadu

Bhathini Vaikuntavasan Pradeep
Bioactive Molecules Laboratory
Department of Microbiology School of Life Sciences
Karpagam University
Coimbatore – 641 021, Tamil Nadu

Binu Thomas
Department of Botany, School of Life Sciences
Bharathiar University, Coimbatore
Tamil Nadu

D. Brindha
Department of Biochemistry
PSG College of Arts and Science
Coimbatore – 641014, Tamil Nadu

D. Patric Raja
Centre for Plant Biotechnology
Department of Plant Biology & Plant Biotechnology
St. Xavier's College (Autonomous)
Palayamkottai, Tamil Nadu

E Natarajan
PG & Research Department of Botany
National College
Trichy – 620 001,Tamil Nadu

E. Meera Nandhini
Department of Biotechnology
University of Madras, Guindy Campus
Chennai – 600 025
Tamil Nadu

Francis Stanly Pradeep
Bioactive Molecules Laboratory
Department of Microbiology
School of Life Sciences Karpagam University
Coimbatore – 641 021,Tamil Nadu

G. Mahendran
Department of Botany
School of Life Sciences
Bharathiar University
Coimbatore – 641 046, Tamil Nadu

G. Vanitha
Department of Botany
Nirmala College for Women
Coimbatore, Tamil Nadu

G.M. Lekshmi
Department of Botany
St. Gregorios College, Kottarakara
Kerala

I. Arul Starlin
Centre for Plant Biotechnology
Department of Plant Biology & Plant Biotechnology
St. Xavier's College (Autonomous) Palayamkottai,
Tamil Nadu

I. Revathy
Centre for Plant Biotechnology
Department of Plant Biology & Plant Biotechnology
St. Xavier's College (Autonomous) Palayamkottai,
Tamil Nadu

Johnson, M.
Centre for Plant Biotechnology
Department of Plant Biology & Plant Biotechnology
St. Xavier's College (Autonomous)
Palayamkottai, Tamil Nadu

K. Harikrishnan
P.G. & Research Department of Botany
National College, Trichy– 620 001
Tamil Nadu

K. Umamaheswari
Department of Biotechnology
University of Madras, Guindy Campus
Chennai
Tamil Nadu

K. Usha
Dept. of Biochemistry, Biotechnology & Bioinformatics
Avinashilingam Institute for Home Science and
Higher Education for Women
Coimbatore – 641 043, Tamil Nadu

K.Kannan
Department of Plant Pathology
Tamil Nadu Agricultural University
Coimbatore – 641 003, Tamil Nadu

K.M. Aravinthan
Dr. Mahalingam Center for Research & Development
Research Department of Biotechnology
NGM College, Pollachi, Tamil Nadu

K.R Vijayalatha
Department of Horticulture
Rice Research Station, Tirurkuppam
Tirur – 602025, Tiruvallur District, Tamil Nadu

K.S. Tamil Selvi
Department of Botany
PSGR Krishnammal College for Women
Peelamedu, Coimbatore, Tamil Nadu

K. Umamaheswari
Department of Biotechnology,
University of Madras, Guindy Campus
Chennai – 600025, Tamil Nadu

Kathirvelu Sambandan
Department of Plant Science, Avvaiyar
Government College for Women, Karaikal – 609 602
U.T of Puducherry

Lakshmanan Nalina
Tamil Nadu Agricultural University
Coimbatore – 641 003, Tamil Nadu

M. Abirami
Department of Biotechnology
University of Madras, Guindy Campus
Chennai, Tamil Nadu

M. Anand
Department of Marine and Coastal Studies
School of Energy Sciences
Madurai Kamaraj University, Madurai
Tamil Nadu

M. Babuselvam
P.G. Department of Microbiology
Dr. Zakir Hussain College
Ilayangudi, Tamil Nadu

M. Dhinakarraj
Department of Biotechnology
PRIST University
Thanjavur, Tamil Nadu

M. Johnson
Centre for Plant Biotechnology
Department of Plant Biology & Plant Biotechnology
St. Xavier's College (Autonomous)
Palayamkottai, Tamil Nadu

M. Ramesh
Department of Biotechnology
Dr. N.G.P Arts and Science College
Coimbatore, Tamil Nadu – 641 048

Mariappan Suganthy
Tamil Nadu Agricultural University
Coimbatore – 641 003
Tamil Nadu

Mrinalini J. Singh
Associate Professor
Department of Botany
Nirmala College for Women
Coimbatore, Tamil Nadu

N. Janakiraman
Centre for Plant Biotechnology
Department of Plant Biology & Plant Biotechnology
St. Xavier's College (Autonomous)
Palayamkottai
Tamil Nadu

N. Ramesh Kannan
PG & Research Department of Botany
National College
Trichy – 620 001
Tamil Nadu

Narayanasamy Dhatchanamoorthy
Department of Plant Science
Avvaiyar Government College for Women
Karaikal – 609 602
U.T o f Puducherry

P. Padma Rao
Drug Standardisation Unit, O.U.B.
32 Road No. 4, Habsiguda
Hyderabad – 500 007, Andhra Pradesh

P. Ranjitha
Department of Biotechnology
University of Madras Guindy Campus
Chennai – 600 025, Tamil Nadu

P. Subramanian
Drug Standardisation Unit O.U.B.
32, Road No. 4, Habsiguda
Hyderabad – 500 007, Andhra Pradesh

P. Sudhakar
Drug Standardisation Unit O.U.B.
32, Road No. 4, Habsiguda,
Hyderabad – 500 007, Andhra Pradesh

P.R. Reddy
Department of Botany
Osmania University
Hyderabad – 500 007, Andhra Pradesh

P.S.Swamy
Department of Plant Science
Madurai Kamaraj University
Madurai – 625 021, Tamil Nadu

Periyaswamy Sakthivel
Tamil Nadu Agricultural University
Coimbatore – 641 003
Tamil Nadu

R. Baskar
Department of Biotechnology
University of Madras, Guindy Campus
Chennai, Tamil Nadu

R. Beena
Department of Biochemistry
PSG College of Arts and Science
Coimbatore – 641 014

R. Gnanasekar
Department of Biotechnology
University of Madras, Guindy Campus
Chennai – 600 025
Tamil Nadu

R. Gomathi
Department of Biochemistry
Biotechnology and Bioinformatics
Avinashilingam Institute for Home Science and Higher Education for Women
Coimbatore – 641 043, Tamil Nadu

R. Rakkimuthu
Dr. Mahalingam Center for Research and Development Research
Department of Biotechnology
NGM College, Pollachi, Tamil Nadu

R. Gnanasekar
Department of Biotechnology
University of Madras, Guindy Campus
Chennai – 600 025
Tamil Nadu

R.M. Vijayakumar
Department of Medicinal and Aromatic Crops
Tamil Nadu Agricultural University
Coimbatore – 641 003
Tamil Nadu

S. Ravikumar
School of Marine Sciences
Deppt .of Oceanography and Coastal Area Studies
Alagappa University, Thondi Campus
Thondi, Ramanathapuram District
Tamil Nadu

S. Selvasubramanian
Department of Veterinary Pharmacology and Toxicology
Madras Veterinary College
Chennai – 600 007
Tamil Nadu

S. Padmavathy
Department of Botany
Nirmala College for Women
Coimbatore
Tamil Nadu

S.A. Ramyabharathi
Department of Plant Pathology
Tamil Nadu Agricultural University
Coimbatore – 641 003
Tamil Nadu

Sentrayaperumal Sundareswaran
Tamil Nadu Agricultural University
Coimbatore – 641 003
Tamil Nadu

T. Renisheya Joy Jeba Malar
Centre for Plant Biotechnology
Department of Plant Biology & Plant Biotechnology
St. Xavier's College (Autonomous) Palayamkottai,
Tamil Nadu

T. Sheshashena Reddy
Drug Standardisation Unit O.U.B.
32, Road No. 4, Habsiguda
Hyderabad – 500 007
Andhra Pradesh

Thangaraj Parimelazhagan
Bioprospecting Laboratory
Department of Botany, Bharathiar University
Coimbatore – 641 046, Tamil Nadu

Thankarajan Sajeesh
Bioprospecting Laboratory
Department of Botany, Bharathiar University
Coimbatore – 641 046
Tamil Nadu

V. Aravindhan
Department of Botany
School of Life Sciences, Bharathiar University
Coimbatore – 641 046
Tamil Nadu

V. Kalaiarasi
Centre for Plant Biotechnology
Department of Plant Biology & Plant Biotechnology
St. Xavier's College (Autonomous)
Palayamkottai, Tamil Nadu

V. Ranganathan
Department of Veterinary Pharmacology and Toxicology
Madras Veterinary College
Chennai – 600 007, Tamil Nadu

V. Sornajana
Department of Biotechnology
University of Madras, Guindy Campus
Chennai – 600 025, Tamil Nadu

V. Logaprabha
Department of Botany
PSGR Krishnammal College for Women
Peelamedu, Coimbatore, Tamil Nadu

V. Mekala
Department of Bioinformatics
Nirmala College for Women
Coimbatore, Tamil Nadu

V. N. Satakopan
Department of Biochemistry
PSG College of Arts and Science
Coimbatore – 641 014, Tamil Nadu

VijayakumarUsha Rani
Bioactive Molecules Laboratory
Department of Microbiology
School of Life Sciences, Karpagam University
Coimbatore – 641 021, Tamil Nadu

1

Phytochemical Analysis and Antimicrobial Properties of *Aegle marmelos* Against Clinically Significant Pathogens

Baskar, R., Abirami, M. and Umamaheswari, K.

1. INTRODUCTION

Medicinal plants are rich source of novel drugs that forms the ingredients in traditional systems of medicine, modern medicines, nutraceuticals, food supplements, folk medicines, pharmaceutical intermediates, bioactive principles and lead compounds in synthetic drugs (Ncube, 2008). In recent years, antibiotic resistance has become a global concern. This drives the need to screen medicinal plants for novel bioactive compounds as they are biodegradable, safe and have fewer side effects (Prusti *et al.*, 2008). Higher plants, as a source for new potential drugs is still largely unexplored and screening of active agents helps to design the lead molecule later in drug development.

Aegle marmelos(L.) Corr., belongs to the family Rutaceae, and is popularly known as Vilvam tree (Tamil) (Gamble, 1935; Mathew, 1983). It is an indigenous tree found in Iran, India, Myanmar, Pakistan, Bangladesh and most of Southeast Asian countries (Singh, 2009). *A. marmelos* has been used from time immemorial in traditional systems of medicine for relieving diarrhoea, dysentery, stomachic, digestive, astringent, peptic ulcer and respiratory infection (Nadkarni, 2000). The fruit, leaves, bark and root are edible and possess medicinal values. The unripe dried fruit used to be astringent, digestive and stomachic, the ripe fruit is aromatic, cooling and is used as laxative. The roots and the bark of the tree are used in the treatment of fever, abdominal pain, the palpitation of the heart, and urinary infections (Kirtikar and Basu, 1935). The bael contains many functional and bioactive compounds such as carotenoids, phenolic compound, alkaloids, coumarins, flavonoids, terpenoids, cardiac glycosides and other antioxidants which may protect against chronic diseases (Singh, 2009). In the present study, we have evaluated the phytochemical analysis and antimicrobial potential of *A. marmelos in vitro* against several clinical pathogens by macro broth dilution technique.

2. MATERIALSAND METHODS

2.1Collection of Plant Material

Fresh leaves and fruits of *A. marmelos* were collected from the field located in Rajapalayam, Tamil Nadu, India. It was identified and authenticated by Dr. Rathinakumar, Department of Botany, Presidency College, Chennai.

2.2 Preparation of Extracts

The leaves and fruits were air dried and powdered. About 25g of powder was packed in a Whatman No.1 filter paper and extracted with ethanol and ethyl acetate solvent in soxhlet extractor for 48 hours. The solvent extracts were concentratedunder reduced pressure and preserved at 4°C in an air tight bottle for further use.

2.3. Microorganisms Tested

The antibacterial activity of the extracts were tested on five clinically significant bacterial species includes *Staphylococcus aureus, Escherichia coli, Vibrio cholerae, Salmonella typhi* and *Proteus mirabilis*.

2.4 Preliminary Phytochemical Analysis

The ethanolic and ethyl acetate extracts of leaves and fruits of *A.marmelos* were subjected to preliminary phytochemical screening to analyze the presence of alkaloids, terpenoids, saponins, tannins, phenolic compounds, flavonoids, cardiac glycosides and steroids (Sofowora, 1982; Trease and Evans, 1989; Adetuyi *et al.*, 2001).

2.5 Antibacterial testing by Macrobroth dilution technique

The test organisms maintained on nutrient agar slants were subcultured on Nutrient agar plates to obtain pure colony. The isolated pure colony were inoculated into Mueller Hinton broth and incubated at 37° C in a shaker at 180 rpm till the concentration of the test organism matched with the 0.5 McFarland standard. Equal volume of the Mueller Hinton broth and extracts were mixed and 20µl of inoculum suspension and the tubes were incubated at 37° C for 24 hours. The lowest concentration of the extracts that inhibited the growth of the bacterial species was noted and MIC (Minimum Inhibitory Concentration) values were determined. Dilutions of ampicillin served as positive control,while tubes without the extract served as control.

2.6. Determination of Minimum Bactericidal Concentration

The minimum bactericidal concentration (MBC) was determined by inoculating each of the dilutions from MIC tubes onto fresh Mueller-Hinton agar plates. The plates were incubated at 37° C for 24 hours and the lowest dilution that killed the bacterial species was recorded as MBC.

3. RESULTS AND DISCUSSION

The MIC and MBC values of the ethanolic and ethyl acetate extracts of leaves and fruits of *A.marmelos* on different bacterial species are shown in Table 1 and 2. The MIC values of both the extracts ranged from 4mg/ml to 16mg/ml. Ethyl acetate extract of leaves of *A. marmels* showed inhibition against *V. cholerae* and *E. coli* at 4mg/ml and for *P. mirabilis* and *S. aureus* at 8mg/ml whereas ethanolic extract of leaves against *V. cholera* were 8mg/ml and for *E. coli* and *P. mirabilis* at 16 mg/ml. The ethyl acetate extract of the fruits showed a MIC value of 4mg/ml against *Proteus mirabilis* and *E. coli,* 8mg/ml against *S.aureus*and *V. cholera.* While the ethanolic extracts inhibited *S.aureus, E. coli* and *P. mirabilis* at 4mg/ml and *V. cholera* at 8mg/ml. The MIC and MBC for ampicillin against the test organism were 16, 0.3, 0.008, 0.4 and 1.0μg/ml, respectively. Phytochemical analysis revealed the presence of alkaloids, cardiac glycosides, terpenoids, saponins, tannin, flavonoids, and steroids in leaves and fruits of *A.marmelos* and is shown in Table 3.

The emergence of antimicrobial resistance among microbes has led to the search for an alternative therapy and identification of newer, novel and safer antimicrobial agents. Plant has been one of the major sources for antimicrobial compounds and provides lead compounds for development of antimicrobial agents. In the present study, the antibacterial activity of the leaves and fruit extracts of *A. marmelos* was evaluated. The ethanolic and ethyl acetate extracts of fruits and leaves of *A.marmelos* showed inhibitory activity against all the bacterial species tested except *S. typhi*. Venkatesan *et al.*, 2009 reported the antibacterial activity of ethanolic extract of leaves of *A. marmelos* against *E. coli, P. aeruginosa, S. aureus* and *B. subtilis* by disc diffusion method and reported MIC value ranging from 0.5 mg/ml to 2.5 mg/ml while in the present study we observed a higher MIC value of 4mg/ml to 16mg/ml. The author has also reported the presence of alkaloids, cardiac glycosides, terpenoids, saponin, tannins, flavonoids, and steroid in ethanolic extract of leaves of *A. marmelos*. Our result of phytochemical analysis was in agreement with the authors findings. Amitpandey *et al.,* 2011 reported the antibacterial activity of ethanolic, methanolic, ethyl acetate and aqueous extracts of leaves, fruits and

Table 1: MIC and MBC of ethanolic and ethyl acetate extracts of leaves of *A.marmelos*

S.No	Extract used	Bacterial pathogens tested and MIC and MBC values in mg/ml									
		Staphylococcus aureus		*Salmonella typhi*		*Vibrio cholerae*		*Proteus mirabilis*		*Escherichia coli*	
		MIC	MBC	MIC	MBC	MIC	MBC	MIC	MBC	MIC	MBC
1.	Ethanolic extract of leaves	NI	NI	NI	NI	8	16	16	16	16	16
2.	Ethyl acetate extract of leaves	8	16	NI	NI	4	8	8	16	4	8
3.	Ampicillin (Positive Control)	16	16	0.31	0.31	0.008	0.008	0.04	0.04	1.0	1.0

NI- Not inhibited

Table 2: MIC and MBC of ethanolic and ethyl acetate extracts of fruits of *A.marmelos*

S.No	Extract used	Bacterial pathogens tested and MIC and MBC values in mg/ml									
		Staphylococcus aureus		*Salmonella typhi*		*Vibrio cholerae*		*Proteus mirabilis*		*Escherichia coli*	
		MIC	MBC	MIC	MBC	MIC	MBC	MIC	MBC	MIC	MBC
1.	Ethanolic extract of fruit	4	8	NI	NI	8	16	4	8	4	8
2.	Ethyl acetate extract of fruit	8	16	NI	NI	8	16	4	8	4	8
3.	Ampicillin (Positive Control)	16	16	0.31	0.31	0.008	0.008	0.04	0.04	1.0	1.0

NI- Not inhibited

peels of *A.marmelos* against various pathogens and recorded the MIC values of 1.98 mg/ml in ethanolic and ethyl acetate extract of fruits against *S. aureus* and 11.90mg/ml methanolic extract against *P. aeruginosa* while in the present study we observed a higher MIC value of 4mg/ml to 8mg/ml in ethanolic and ethyl acetate extracts of fruits against *S.aureus*.The antibacterial activity testing revealed that ethyl acetate extracts of both leaves and fruits of *A.marmelos* showed slightly higher inhibitory activity than ethanolic extracts. Testing the extracts with larger number of pathogen and phytochemical analysis would help to identify active principles responsible for antimicrobial activity.

Table 3: Phytochemical analysis of leaves and fruit of *A. marmelos*

Phytochemicals Tested	Ethanolic extract of leaves	Ethyl acetate extract of leaves	Ethanolic extract of fruits	Ethyl acetate extract of fruits
Alkaloids	+	+	+	+
Saponins	+	+	-	-
Phenolic compounds	-	-	+	+
Steroids	+	+	+	+
Flavonoids	+	+	+	+
Tannins	+	+	+	+
Cardiac glycosides	+	+	-	-
Terpenoids	+	+	+	+

4. CONCLUSION

From the results of preliminary phytochemical screening and antibacterial activity of two solvents (ethyl acetate, ethanol) used in this study, ethyl acetate was exhibited best antibacterial activity. Among the two plant parts used for testing in this study, leaf of *A.marmelos* was considered as the most effective. Because itexhibited good antibacterial activity against all pathogens tested may be due to the presence of alkaloids, phenols, flavanoids, and Cardiac glycosides. The inability of leaf and fruit extracts of *A. marmelos* to demonstrate any visible activity against *S. typhi* may probably be due to the low concentration of the extracts.

In this endeavor, traditional herbal medicines must perforce be granted the benefits of modern science and technology to serves further global needs. The drugs derived from herbs may have the possibility of using in medicine because of its good antibacterial potentiality. Further research in this pursuit, focusing on the isolation of individual compounds and finally subjecting to clinical trials promises to open new avenues in the use of plants for therapeutic purpose.

REFERENCES

Adetuyi, A.O., Popoola, A.V., 2001. Extraction and dyes ability potential studies of the colourant inzanthoxylumzanthoxyloides plant on cotton fabric. *Journal of Science Engineering Technology*.8 (2): 3291-3299.

Amit Pandey, Rashmi Mishra, 2011. Antibacterial properties of *Aegle marmelos*leaves, fruits and peels against various pathogens. *JPBMS,* 13 (13).

Gamble, J.S., 1935.Flora of the Presidency of Madras.Adlard and Son's Ltd, London, UK

Kirtikar K.R, Basu S.D, 1935. Indian Medicinal Plants. Vol. III. Lalit Mohan Basu, Allahabad, India

Matthew, K.M., 1983.The Flora of Tamil Nadu Carnatic. The Rapinat Herbarium, St Joseph's College, Tiruchirapalli, India

Ncube, N., Afolayan, S.A.J., Okoh, A.I., 2008. Assessment techniques of antimicrobial properties of natural compounds of plant origin: current methods and future trends. *Afr. J. Biotechnol.* 7(12): 1797-1806

Nadkarni, A.K., 2000, Indian materiamedica, IIIrdEdn. Popular press, Mumbai, India, 45-49

Prusti, A., Mishra, S.R., Sahoo, S., Mishra., S.K., 2008. Antibacterial Activity of Some Indian Medicinal Plants. *Ethnobotanical Leaflets*.12: 227-230.

Sofowora, A., 1982. *Medicinal plants and Traditional medicine in West Africa*, John Wiley and Sons, pp. 256, New York.

Singh, R., Singh H. Rao., 2008. Hepatoprotective effect of the pulp/seed of *Aegle marmelos* correa ex Roxb against carbon tetrachloride induced liver damage in rats. *International Journal of Green Pharmacy*, 232

Trease, G.E., Evans, W.C., 1989. Pharmacognosy 11th Edn.BrailliarTiridal Can Macmillian publishers

Venkatesan, D., Karrunakarn, C.M., Kumar, S.S., Swamy, P.T.P., 2009. *Ethnobotanical Leaflets*, 13, 1362-1372.

2

Screening of *Gloriosa superba* Ecotypes Against leaf Blight Disease

B. Meena, K. Kannan, S.A. Ramyabharathi and R.M. Vijayakumar

1. INTRODUCTION

Gloriosa superba L. (Liliaceae) is a striking tuberous climbing plant with brilliant wavy-edged yellow and red petaled flowers. Glory lilies are grown commercially for a chemical compound, colchicine. The annual foreign exchange through the export of glory lily seeds is estimated to be Rs.100 crores (Alice and Sundravadana, 2012). It is used as anthemirtic, alexiteric and useful in ulcers, piles, abdominal pains and itching. It has abortifacient, anti-pyretic, anti-inflammatory and anti-leprotic properties (Gupta *et al.,* 2005). Leaf blight disease was first noticed in *G. Superba* in West Bengal, India in 2004. The leaf blight disease was observed during June to September. Early symptoms appeared as small, circular to oval, light brownish spots, 2-6 per leaf, scattered at the tip, margin and midrib of the leaves. Subsequently the spots enlarged and usually developed into concentric rings. At an advanced stage, the spots became dark brown to blackish in colour, gradually coalesced and became irregular in shape, then the affected leaf blighted completely. The causal organism was found to be a fungus namely *Alternaria alternata* (Fr.) Keissler. The pathogen is greatly influenced by weather with the highest disease incidence reported in wet seasons and in areas with relatively high rainfall (Meena *et al.,* 2010). The growth of *A. alternata* was maximum in pH range of 6-6.5 and temperature range of 25-30°C (Hubballi *et al.,* 2010).

Symptoms appear in all stages. Initially, small, pale to brown, irregular or round spots appear on the leaves. Each spot has a central necrotic lesion with concentric rings. In advanced stage, several spots coalesce together to form large blighted areas. The disease also occurs on the petiole and flowers. The pollination gets affected due to leaf blight pathogen infection and consequently the yield is drastically reduced to the tune of 30 to 40%. Management of disease through fungicides alone leads to cause soil residual problem and health hazards, besides involving higher input cost. Host resistance is one of the important components of integrated disease management.

2. MATERIALS AND METHODS

The tubers from eighteen ecotypes of *G. superba* were collected from different places of India and sown under field conditions. The different ecotypes of *(G. superba)* were screened against leaf blight disease under natural field conditions. The leaf blight disease intensity was assessed on 90 days after planting using 0-9 disease rating scale as described by Pawelec *et al.*, 2006 [0 : No visible disease damage; 1 : <5% leaf area damaged; 3 : 5-20% leaf area damaged; 5 : 20-40% leaf area damaged; 7 : 40-60% leaf area damaged; 9 : severe defoliation].

Per cent disease index (PDI) for each ecotype was calculated using the formula

$$\text{PDI} = \frac{\text{Sum of numerical ratings}}{\text{Number of leaves examined x Maximum grade available in the score chart}} \times 100$$

3. RESULTS AND DISCUSSION

Breeding for disease resistance requires efficient, low-cost and rapid screening techniques (Foolad *et al.*, 2000). Eighteen ecotypes of *G. superba* were collected from different places of India and screened against leaf blight disease under natural field conditions. All the *G. superba* ecotypes were found to be infected with leaf blight disease. The lowest leaf blight disease intensity was observed in GS03, GS04, GS12 and GS14 ecotypes. The highest leaf blight disease intensity was recorded in GS08 and GS18 ecotypes (Table 1). The promising ecotypes can be utilized as resistant donors in the breeding programme. Several sources of tolerance against Alternaria blight have been reported (Gupta *et al.*, 2001).

Resistance to Alternaria blight in mustard has been found to be associated with leaf enzymes associated with the phenolic pathway, e.g., polyphenol oxidase, peroxidase, catalase, higher leaf sugar contents (Singh *et al.*, 1999) and high deposits of leaf epicuticular wax forming a hydrophobic coating to reduce the adherence of water-borne inoculum, conidia germination and germtube formation (Saharan, 1992). Krishnia *et al.* (2000) indicated the effect of additive genes or polygene or cluster gene on mechanism of tolerance to Alternaria blight. Since resistance to Alternaria blight is found to be governed by additive or polygenes, breeding for resistance to these diseases could involve pyramiding of minor genes affecting the disease to provide additive/polygene resistance against the disease, introgression of genes from material found resistant, reciprocal recurrent selection or diallel selective mating (Krishnia *et al.*, 2000).

Table 1: Screening of Gloriosa ecotypes against leaf blight disease under field conditions

Ecotypes	Alternaria leaf blight (PDI)
GS01	18.2
GS02	16.3
GS03	11.2
GS04	13.4
GS05	18.4
GS06	20.6
GS07	22.5
GS08	26.8
GS09	24.2
GS010	25.3
GS011	21.4
GS012	12.9
GS013	25.6
GS014	13.1
GS015	22.4
GS016	25.8
GS017	24.4
GS018	26.2
Loacal check	34.1

4. CONCLUSION

The study indicates the scope of developing resistant variety to leaf blight disease by using the resistant sources now available for breeding programme.

REFERENCES

Alice, D., Sundravadana, S., 2012. Effects of biocontrol agents and plant products on *Macrophomina phaseolina* and colchicine content in *Gloriosa superba*. *Plant Protect. Sci* 48, 110-115.

Foolad, M.R., Ntahimpera, N., Christ, B.J., Lin, G.Y., 2000. Comparison of field, greenhouse, and detached-leaflet evaluations of tomato germplasm for early blight resistance. *Plant Disease* 84, 967-972.

Gupta, K., Saharan, G.S., Singh, D., 2001. Sources of resistance in Indian mustard against white rust and Alternaria blight. *Cruciferae Newslett* 23, 59-60.

Gupta, L.M., Rana, R.C., Raina, R., Gupta, M., 2005. Colchicine content in *Gloriosa superba* L. *Research SKUAST Journal* 4, 238-241.

Hubballi, M., Nakkeeran, S., Raguchander, T., Anand, T., Samiyappan, R., 2010. Effect of environmental conditions on growth of *Alternaria alternata* causing leaf blight of noni. World Journal of *Agricultural Sciences* 6, 171-177.

Krishnia, S.K., Saharan, G.S., Singh, D., 2000. Genetic variation for multiple disease resistance in the families of interspecific cross of *Brassica juncea* x *B. carinata. Cruciferae Newsletter* 22, 51-53.

Meena, P.D., Awasthi, R.P., Chattopadhyay, C., Kolte, S.J., Arvind Kumar., 2010. Alternaria blight : a chronic disease in rapeseed-mustard. *Journal of Oilseed Brassica* 1, 1-11.

Pawelec, A., Dubourg, C., Briard, M., 2006. Evaluation of carrot resistance to alternaria leaf blight in controlled environments. *Plant Pathology* 55, 68-72.

Saharan, G.S., 1992. Disease resistance. In : Breeding Oilseed Brassicas. K.S. Labana, S.S. Banga, S.K. Banga (Eds). Narosa Pub. House, New Delhi, India, pp. 181-205.

Singh, D.N., Singh, N.K., Srivastava, S., 1999. Biochemical and morphological characters in relation to Alternaria blight resistance in repeseed – mustard. *Annals. Agril. Res* 20, 472-477.

3

Medicinal Plants Sold in the Velliangiri Hills, the Southern Western Ghats of Coimbatore District, Tamil Nadu

Aravindhan, V., Mahendran, G. and Rajendran, A.

1. INTRODUCTION

Plants are reliable sources for the treatment of diseases in different parts of the world (Eisenberb *et al.*, 1993; Cowan, 1999; Hostettmann *et al.*, 2000) and they have been unique sources of medicines and constituted the most common use of biodiversity (Hamilton, 2004; Hiremath and Taranath, 2010). Their use contributes significantly to primary healthcare delivery (Holetz *et al.*, 2002) as they are regarded as invaluable sources of pharmaceutical products (Olalde, 2005).

Traditional medicine, with medicinal plants as their most important component, are sold in market places or prescribed by traditional healers in their homes (Von Maydell, 1996). As a result of this strong dependence on plants as medicines, many ethnopharmacological studies have been conducted to determine their safety, efficiency and discovery of new active principles from them. Only, 1,100 plants have so far been thoroughly examined for their medicinal practices in the world (Dobson, 1995).

Herbal medicine have been in use since time immemorial and are still the mainstay mainly in the developing countries because of their better cultural acceptability, better compatibility with the human body and lesser side effects (Kamraj, 2000). World Health Organization (WHO) has reported that over 80% of the world's population relies on traditional form of medicine, largely plant based to meet the healthcare needs. Out of *ca* 17,500 flowering plant species found in India, over 16,000 are used in traditional medicinal system (Anon., 1993).

Although the use of medicinal plants for subsistence, home remedies, trade and alleviating human suffering (Kunwar *et al.*, 2006) plays important roles in the lives of rural people (Ahmad, 2003), the non-sustainable collection methods have caused threat from harvesting and many valuable medicinal herbs are becoming rare due to their continuous utilization (Swe and Win, 2005) and over-exploitation for commercial purposes (Tabuti *et al.*, 2003; Kamatenesi-Mugisha and Oryem-Origa, 2005).

The present study was conducted to collect information on medicinal plants sold in the local market at Velliangiri Hills of Coimbatore district by the tribe *Irulars*. They collect and sell the medicinal plants during February – June and the income earned during the period supplements the income generated from agriculture. People have a religious belief that, plants in Velliangiri Hills have got medicinal properties. So, they buy medicinal plants from the local people/tribes and market place represents an important centre of ethnobotanical information. Thus, detailed ethnobotanical studies on medicinal plants sold by the *Irular* tribes were undertaken in the study area.

1.1. Study area

Velliangiri Hills, a major hill range of Southern Western Ghats, situated about 20 km from Coimbatore town towards Siruvani and it forms part of Nilgiri Biosphere Reserve (Map 1). It lies between the longitude 6p - 402 and 7p - 102 E and latitude 10p - 552 and 11p - 102 N and an altitude ranging from 500-1800 m above MSL. The area is floristically very rich and socio-religiously important since a famous temple called Velliangiri Andavar temple also known as *Thenkailayam* (South Kailash) is located at the peak of the hills.

The vegetation of hills ranges from scrub forests at foothill to evergreen forests. The forest types met in this area are tropical thorn forests, tropical dry deciduous forests, tropical wet evergreen forests, grasslands and a small patch of montane shola type of forest is also seen at the top. The soil type is red, loamy, acidic and ferruginous. The average rainfall in the hills is *ca* 3500 mm at foot hills and 4500 mm at the peak per year. The amount of rainfall increases with the increase of altitude.

2. MATERIALS AND METHODS

Field surveys were conducted between 2011 and 2012 to collect information on medicinal plants sold in the local market at Vellingiri Hills (Fig. 1). Information on local name of the plant, part sold, uses, locality of collection and uses were gathered from the shop owners (Fig. 2). Plant specimens were collected and identified with the help of local floras (Gamble, 1957; Matthew, 1983). The voucher specimens were deposited at the Herbarium of Department of Botany, Bharathiar University, Coimbatore, Tamil Nadu for further reference.

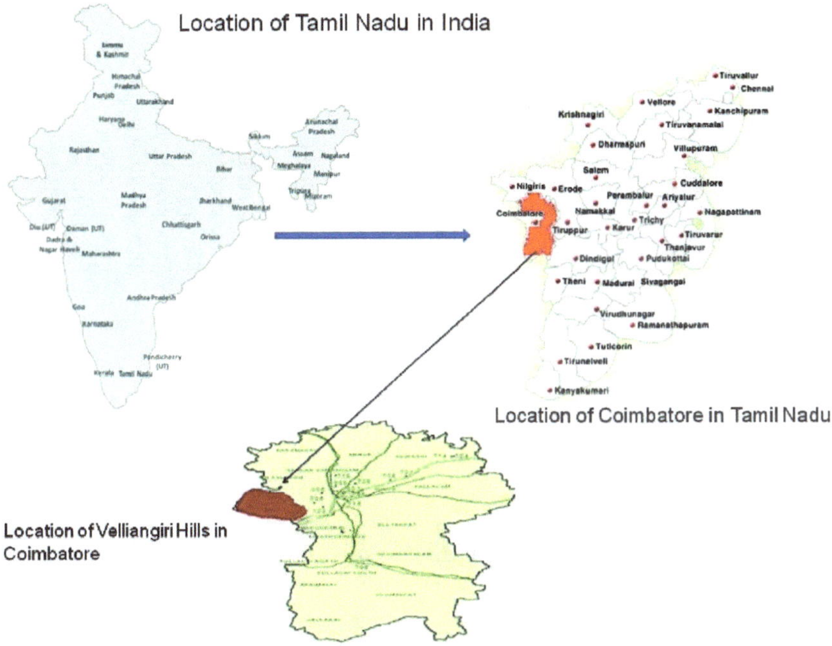

3. RESULTS AND DISCUSSION

A significant amount of medicinal plants is collected and sold locally by the tribe *Irulars* without entering the major market places. They collect and sell medicinal plants in the local market at Poondi, the foothills of Velliangiri during February – June of every year. A total of 42- medicinal plant species belonging to 27- families were recorded from the market place (Table -1). Among the 42- plant species, 15- were herbs, 11- trees, 8- shrubs and 7- climber (Fig. 3).

Most of the medicinal plants sold in the market were collected from the wild while species such as *Acorus calamus*, *Curcuma aromatic*, *Vetiveria zizanioides* and *Zingiber officinale* were procured from nearby major market places such as Boluvampatti, Semmedu, etc. Mostly young men collect these plants and give to the women and elders, who in turn sell them during festival season. The study observed that, most of the medicinal plants were sold as fresh while some of the medicinal plants *Viz*., *Andrographis paniculata*, *Euphorbia rothiana*, *Terminalia chebula*, *T. bellerica* and *Todallia asiatica* were sold after drying.

Among the 42 species, 27% for reproductive parts (flower/fruit/seed), 33% were sold for their underground parts (roots/rhizome), leaves 20%, stem 10%, bark

Table 1: List of medicinal plants sold by tribe *Irulars* in the Velliangiri Hills of Coimbatore District, Tamil Nadu

Species	Local Name	Family	Habitat	Part used	Period of collection	Uses
Acorus calamus L.	Vasambu	Araceae	Shrub	Rhizome	Throughout the year	Cough
Aloe vera (L.) Burm. f.	Sothukathalai	Liliaceae	Herb	Whole plant	Throughout the year	Skin diseases
Andrographis paniculata (Burm. f.) Wall. *ex* Nees	Siriyanangai	Acanthaceae	Shrub	Stem, Leaves	February to June	Stomachache
Angiopteris evecta (G.Forst.) Hoffm.	Alangil	Angiopteridaceae	Herb	Leaves	Throughout the year	Skin diseases
Anisochilus carnosus (L.f.) Wall.	Kalthamarai	Lamiaceae	Shrub	Leaves	Throughout the year	Cough and Cold
Asparagus racemosus Willd.	Thannervittan kilangu	Liliaceae	Climber	Tuber	Throughout the year	Indigestion and sexual impotency
Begonia malabarica Lam.	Rathasoori	Begoniaceae	Herb	Stem	Throughout the year	Skin diseases
Bulbophyllum acutiflorum A. Rich.	Ponmanjal	Orchidaceae	Herb	Bulb	February to June	Health tonic
Bulbophyllum neilgherrense Wight.	Thirowpathi manjal	Orchidaceae	Herb	Whole plant	Throughout the year	Skin diseases and Stomach problems
Canarium strictum Roxb.	Kungilium	Burseraceae	Tree	Bark	Throughout the year	Cough and Fever
Caralluma attenuata Wight	Kattukathalai	Asclepiadaceae	Shrub	Stem	February to June	Diabetes and inflammations
Coelogyne nervosa A. Rich.	Seethai manjal	Orchidaceae	Herb	Tuber	Throughout the year	Health tonic
Curcuma aromatic Salish	Kasthurimanjal	Zinigiberaceae	Shrub	Rhizome	February to June	Skin diseases
Dendrobium herbaceum Lindl.	Yanaimirati	Orchidaceae	Herb	Whole plant	February to June	Earache
Diplocyclos palmatus (L.) C. Jeffrey	Sivalingakai	Convolvulaceae	Climber	Leaves	Throughout the year	Rheumatic pain
Eria mysorensis Lindl.	Ivirali	Orchidaceae	Herb	Tubers	February to June	Blisters and Wounds
Euphorbia rothiana Spreng.	Ponnavarai	Euphorbiaceae	Herb	Leaves	Throughout the year	Jaundice
Fagraea ceilanica Thunb.	Kandankathiri	Loganiaceae	Tree	Fruit	Throughout the year	Skin diseases
Ficus racemosa L.	Atthipalam	Moraceae	Tree	Fruit	February to June	Kidney disorders
Garcinia resinifera Roth.	Kambipisin	Rubiaceae	Tree	Resin	February to June	Cough and Fever
Gymnema sylvestre (Retz.) R. Br. *ex* Schult.	Sarkaraikolli	Asclepiadaceae	Climber	Leaves	February to June	Diabetes

(Contd.)

Species	Local Name	Family	Habitat	Part used	Period of collection	Uses
Helicteres isora L.	Valamupuri	Sterculiaceae	Shrub	Seed	Throughout the year	Dysentery
Hemidesmus indicus (L.) R. Br.	Nannari	Asclepiadaceae	Herb	Root	February to June	Health tonic and Urinary diseases
Impatiens parasitica Bedd.	Thelkadiver	Balsaminaceae	Herb	Root	February to June	Scorpion bites
Medinella beddomei Clarke	Neerkodiver	Melastomataceae	Herb	Root	February to June	Insect bites
Myristica dactyloides Gaertn.	Jathikai	Myristicaceae	Tree	Fruit	Throughout the year	Stomachache
Naringi crenulata (Roxb.) Nicolson	Mahavilvam	Rutaceae	Climber	Fruits	February to June	Dysentery and vomiting
Notonia grandiflora Wall. *ex* DC.	Nilampirandai	Asteraceae	Shrub	Stem	February to June	Skin diseases
Oberonia wightiana Lindl.	Senkathalai	Orchidaceae	Herb	Whole plant	February to June	Skin diseases
Peperomia tetraphylla (G. Forst.) . Hook. & Arn.	Kalporumi	Piperaceae	Herb	Whole plant	February to June	Kidney disorders
Phyllanthus emblica L.	Nellikai	Euphorbiaceae	Tree	Fruits	Throughout the year	Diarrhoea and dysentery
Piper hymenophyllum Miq.	Kattu thipli	Piperaceae	Climber	Leaves	February to June	Inflammations
Piper longum L.	Thippili	Piperaceae	Climber	Root	Throughout the year	Cough and Bronchitis
Pterocarpus marsupium Roxb.	Vengai	Fabaceae	Tree	Resin	Throughout the year	Dysentery
Schefflera racemosa (Wight) Harms	Peimiratti	Araliaceae	Tree	Flowers	February to June	Toothache
Strychnos nux-vomica L.	Ettikai	Loganiaceae	Tree	Seed	February to June	Fever
Terminalia bellirica (Gaertn.) Roxb.	Thanikkai	Combretaceae	Tree	Fruit	Throughout the year	Stomachache
Terminalia chebula Retz.	Kadukka	Combretaceae	Tree	Fruit	Throughout the year	Cough and Stomach ulcer
Toddalia asiatica (L.) Lam.	Milakaranai	Rutaceae	Climber	Leaves and Root bark	February to June	Health tonic and Stomach ulcer
Urginea indica Kunth	Kattuvengayam	Liliaceae	Herb	Tubers	February to June	Cough and Bronchitis
Vetiveria zizanioides L.	Vetiver	Poaceae	Shrub	Root	February to June	Fever

Fig. 1: Tribal local market at Velliangiri Hills

Fig. 2: Interaction awith shop owners

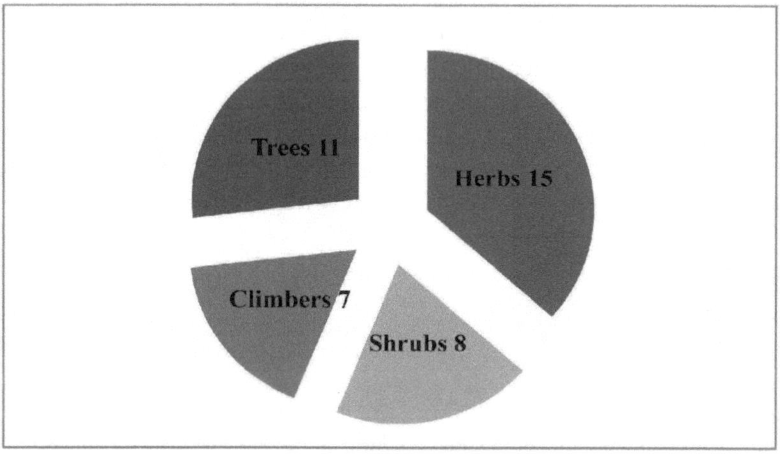

Fig. 3: Analysis of life-forms in the study

5% and resin 5% (Fig. 4). As compared to the collection of aerial parts, the collection of underground parts of a plant will be more destructive and it will ultimately affect the wild population. Destructive collection of medicinal plants seems to be a common practice in the study area. However, quantitative data on distribution, abundance, regeneration and population structure is required for all medicinal plants to assess the sustainability of utilization of medicinal plants in this area.

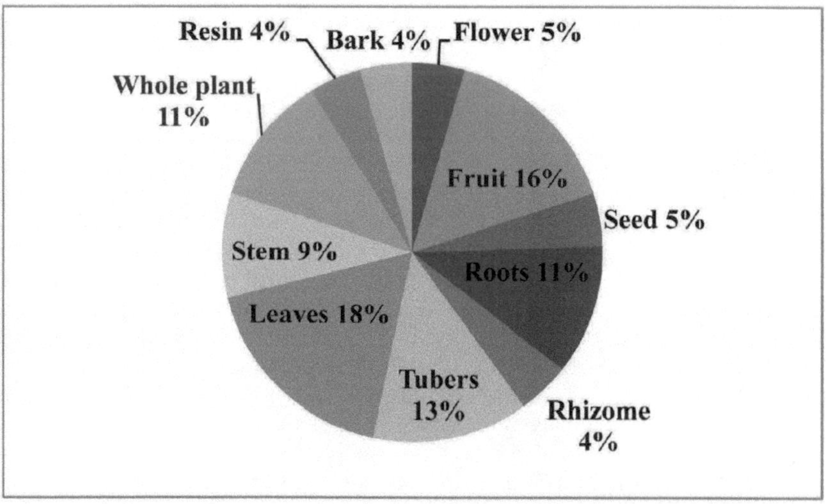

Fig. 4: Analysis of utilization of different plant parts in the study

4. CONCLUSION

The present study concluded that, to enlighten the negative impacts of over-exploitation and to ensure sustainable extraction, the *Irular* tribes who are involved in the collection of medicinal plants should be given sufficient training in the scientific way of collection and sustainable utilization of bioresources. As a result, a sustained yield from the forests can be achieved by diminishing the collection of medicinal plants.

ACKNOWLEDGEMENTS

Authors are thankful to the authorities of University Grants Commission (UGC), New Delhi for providing financial assistance under Major Research Project (F. No. 39-422/2010 (SR) dated 7th January 2011). We are also thankful to Joint Director, Botanical Survey of India, Southern Regional Centre, Coimbatore for permitting to consult the herbarium and library.

REFERENCES

Ahmad, M., 2003. Ethnobotanical and taxonomic studies of economically important plants of Tehsil Attock. M. Phil. Thesis, Quaid-e-Azam University, Islamabad, Pakistan. pp. 205–207.

Anonymous, 1993. *Conservation of Medicinal Plants of India*. Botanical Survey of India (BSI), Calcutta.

Cowan, M.M., 1999. Plant products as antimicrobial agents. *Clin. Microbiol. Rev.* 12:564–582.

Dobson, A., 1995. Biodiversity Human health Trends. *Ecol. Evol.* 10 (10): 390.

Eisenberg, D.M., Kessier, R.C., Foster, C., Norlock, F.E., Calkins, D.R., Delbanco, T.L., 1993. Unconventional medicine in the United States: Prevalence, costs and patterns of use. *The New Engl. J. Med.* 328: 246 – 252.

Gamble, J.S., 1957. *Flora of the Presidency of Madras*. (Reprinted edition). Vols. I–III. Calcutta.

Hamilton, A.C., 2004. Medicinal plants, conservation and livelihoods. *Biodiver. Conser.* 13:1477–1517.

Hiremath, V.T., Taranath, T.C., 2010. Traditional phytotheraphy for snake bites by tribes of Chitradurga District, Karnataka, India. *Ethnobot. Leafl.* 14:120–125.

Holetz, F.B., Pessini, G.L., Sanches, N.R. Cortez, D.A.G., Nakamura, C.V., Dias, F.B.P., 2002. Screening of some plants used in the Brazilian folk medicine for the treatment of infectious diseases. *Mem. Inst. Oswaldo Cruz* 97: 1027 – 1031.

Hostettmann, K., Marston, A., Ndojoko, K., Wolfender, J., 2000. The potential of Africa Plants as a source of drug. *Curr. Org. Chem.* 4: 973 – 1010.

Kametenesi-Mugisha, M., Oryem-Origa, H., 2005. Traditional herbal remedies used in the management of sexual impotence and erectile dysfunction in western Uganda. *Afr. Health Sci.* 5 (1): 40–49.

Kamraj, V.P., 2000. Herbal medicine. *Curr. Sci.* 78 (1): 35–39.

Kunwar, R.M., Nepal, B.K., Kshhetri, H.B., Rai, S.K., Bussman, R.W., 2006. Ethnomedicine in Himalayas: a case study from Dolpa, Humla, Jumla and Mustang districts of Nepal. *J. Ethnobiol. Ethnomed.*, 2: 27.

Matthew, K.M., 1983. *The flora of Tamilnadu Carnatic*. The Repinat Herbarium, Tiruchirapalli, Tamil Nadu, India.

Olalde, R.J.A., 2005. The systematic theory of living systems and relevance to CAM. Part I: The theory. *Evid. Based Comp. Alt. Med.* 2: 13–18.

Swe, T. Win, S., 2005. Herbal gardens and cultivation of medicinal plants in Myanmar regional consultation of development of traditional medicine in the South East Asia region, Department of Traditional Medicine, Ministry of Health, Myanmar, Pyongyang, DPR Korea, 22–24 June, 2005. World Health Organization (Regional office for South East Asia).

Tabuti, J.R.S., Dhillion, S.S., Lye, K.A., 2003. Traditional medicine in Bulamogi county, Uganda: its practitioners, users and viability. *J. Ethnopharmacol.* 85: 119–129.

Von Maydell, H.T., 1996. *Trees and Shrubs of the Sahel*. Josef Margraf, Weikersheim, Germany. pp. 562.

4

Evaluation of Antibacterial Potential of *Phyllanthus emblica* L. Against Some Selected Fish Pathogens

Renisheya Joy Jeba Malar T., Johnson M., Arul Starlin, I. and Revathy I.

1. Introduction

Medicinal plants are a basis of huge economic value all over the world. The use of plants and plant products as remedies could be sketched as far back as the beginning of human civilization. The most primitive medicinal use of plants in Hindu culture is mentioned in "Rigveda", which is believed to have been written between 4500 - 1600 B.C. and is theoretical to be the oldest arsenal of human knowledge (Dhale and Mogle, 2011). Ayurveda is the base of curative science of Hindu culture, in its eight dissection deals with specific properties of drugs and different aspects of science of life and the art of healing (Rastogi and Mehrotra, 2002). Nature has presented on us an affluent botanical wealth and a large number of varied types of plants nurture in different parts of the country. Herbal medicine is still the basis of about 75 - 80% of the whole population. The major part of traditional therapy occupies the use of plant extract and their active components (Akerele, 1993). Secondary metabolites are green medicines present in plants and plant parts that act as defense system against disease inducing organisms or more accurately to protect against disease due to the physiological and biochemical mechanism failure (Alarcon-Aguilar *et al.,* 2002).

Phyllanthus emblica L. is a wealthy resource of vitamin C. It have fatty acids like linolenic (8.8%), linoleic (44.0%), oleic (28.4%), stearic (2.15%), palmitic (3.0%)

and myristic acid (1.0%) (Thakur *et al.,* 1994).The fruits, leaves and barks of *P. emblica* are prosperous in tannins. *P. emblica* has gained attention in the potential treatment for human bone disorders (Piva *et al.,* 2009) as well as diabetes patients (Kusirisin *et al.,* 2009). The ellagic acid and lupeol are present in root and leucodelphinidin in bark of *P. emblica*. The leaves and seeds decoctions of *P. emblica* is used in the treatment of diabetes mellitus (Treadway and Linda., 1994). In addition, *P. emblica* possess ethnobotanical application as anti-inflammatory and anti-pyretic agents by the rural population and used to cure scurvy, cancer and heart diseases (Summanen, 1999). The *in vitro* studies proved the anti-neutrophilic activity and anti-platelet properties of *P. emblica* leaves. In addition, the *P. emblica* leaves extracts also possess several pharmacological properties like anti-viral (HIV, AIDS, HERPES VIRUS, CMV) antimutagenic, anti-allergic and anti-bacterial (Khopde *et al.,* 2000). *P. emblica* contains various secondary metabolites with different ratio (Calixto *et al.,* 1998). Malays use the decoction of *P. emblica* leaves to treat fever (Burkill, 1966). In Indonesia, the pulp of the fruit is smeard on the head to dispel headache and dizziness caused by excessive heat (Perry, 1980). Dhale and Mogle (2011) optimzed the extraction procedure and revealed the phytochemical constituents of the dried powdered plant fruits and leaves. Raghu and Ravindra (2010) studied the antibacterial potential of *P. emblica* with 5, 10 and 20 mg/ml diethylether, cloroform and methanolic extracts against *S. aureus* and *K. pneumoniae*. Saeed and Tariq (2007) reported the antibacterial potential of *P. emblica* against some selected human pathogens *Escherichia coli*, *K. ozaenae*, *Klebsiella pneumoniae*, *Proteus mirabilis*, *Pseudomonas aeruginosa*, *S. paratyphi* A, *S. paratyphi* B and *Serratia marcescens* Lakshmi and Venkataraju (2012) studied the phytoprofile of the *Phyllanthus* speceis from Andhra Pradesh. But there is no report on bio-efficacy of *P. emblica* agaisnt the fish pathogens. With this background, the present study was intended to study the phytochemical properties and antibacterial efficacy of *P. emblica* against the fish pathogens.

2. MATERIALS AND METHODS

2.1 Collection of Plant Material

The whole plants of *Phyllanthus emblica* L. was collected from Nagercoil, Kanyakumari district, Tamil Nadu, India. The fresh materials were shade dried and powdered using the tissue blender. 10 g air dried powder of the plant samples were extracted by cold extraction with 60 ml of hexane, petroleum ether, methanolic and aqueous for 72 h under dark with intermittent shaking. After incubation, the solution was filtered through filter paper and the filtrate was collected and stored at 4°C for the biological activity (crude extracts).

2.2 Preliminary Phytochemical Analysis

Preliminary phytochemical screening of different extracts of *P. emblica* was carried out by the standard method (Harborne, 1998) to know the presence of steroids, alkaloids, saponins, tannins, amino acids, phenolics, flavonoids and anthraquinones.

2.2 Antibacterial Assay

The bacteria were isolated from the diseased fish, identified and confirmed by conventional microbiology procedure (Hardi *et al.,* 2004; Mukherjee, 2004). For bio-efficacy studies, the stock cultures of *Klebsiella pneumoniae, Bacillus subtilis, Escherichia coli* and *Pseudomonas aeruginosa* were grown in nutrient broth at 30°C and were sub-cultured and maintained in nutrient broth at 4°C. The crude methanolic extract of *P. emblica* was used for bio-efficacy against the isolated fish pathogens viz., *K. pneumoniae, B. subtilis, E. coli* and *P. aeruginosa*. Anti-bacterial study was carried out by disc diffusion method against the isolated pathogens. Sterile discs with 6 mm diameter were loaded with 25, 50, 75 and 100 µg (w/v) of plant methanolic extracts and inoculated onto the sterile medium with the pathogens. The plates were incubated at 37°C for 18-24 h. Antibacterial activity was evaluated by measuring the diameter of zone of inhibition. The antibiotic chlormphenicol (30 µg) was served as positive control and methanol (95%) was treated as negative control. Each and every experiment was repeated thrice and the results were recorded.

3. RESULTS AND DISCUSSION

3.1 Phytochemical Evaluation

The qualitative phytochemical screening of *P. emblica* revealed the presence of alkaloids, steroids, phenolics, tannins, saponins, flavonoids, anthraquinones and amino acids with varied degree (Table 1). In the present investigation, preliminary phytochemical screening was carried out using four different extracts viz., petroleum ether, hexane, methanolic and aqueous. Among the four different extracts, methanolic extracts of *P. emblica* showed the maximum occurrence of secondary metabolites viz., alkaloids, saponins, tannins, phenolics and flavonoids. The aqueous extract of *P.emblica* showed the presence of tannins, alkaloids, phenolics and flavonoids. Petroleum ether and hexane extract of *P. emblica* illustrated the unique existence of steroids and flavonoids. Lakshmi and Venkataraju (2011) reported the methanolic extracts of *P. emblica* possessing alkaloids, coumarins, dihydro chalcones, flavonoids, fatty acids, glycosides, gallic tannins, phenolics. Petroleum ether and aqueous extract of *P. emblica* illustrated the presence of alkaloids and carotenoids. In the present investigation also the methanolic extract of *P. emblica* expressed maximum number of bioactive compounds. Our results were directly coinicided with Lakshmi and Venkata raju observations. Dhale *et al.,* (2011) revealed the phytprofile of *P. emblica* in three different extracts viz., petroleum ether, chloroform and ethanol. Among the tested three extracts, ethanolic extract showed the maximum expression of the bioactive compounds viz., alkaloids, oil and fats, glyceroids, carbohydrates, phenolics, tannins, lignin, saponins, flavonoids and terpenoids. Petroleum ether and chloroform extracts conveyed the presence of alkaloids, oil and fats, phenolics, tannins, lignins and saponins. But in the present study, petroleum ether extract of *P. emblica* failed to show the presence of alkaloids, phenolics, saponins and tannins and it was contrary to Dhale *et al.,* observations. The phytochemical profile varaitions may be due to the ecological or gentical variations. Hence, methanolic extract was screened for further antibacterial activity.

Table 1: Phytochemical Analysis of *Phyllanthus emblica.*

Phytochemical constituents	Solvents			
	Hexane	Petroleum ether	Methanol	Aqueous
Steroid	-	+	-	-
Alkaloid	-	-	+	+
Saponin	-	-	+	-
Anthraquinone	-	-	-	-
Aminoacid	-	-	-	-
Phenolics	-	-	+	+
Tannin	-	-	+	+
Flavonoids	+	-	+	+

+) Presence of chemical compound (-) Absence of chemical compound

3.2 Antibacterial activity

Different concentrations (25to100 µg/ml) of *P. emblica* methanolic extract was examined for the antibacterial activity against the fish pathogens and the results were depicted (Table 2). With reference to the concentrations the zone of inhibition against the fish pathogens was varied. The 100 µg/ml of *P. emblica* methanolic extracts showed the maximum zone of inhibition with 15 mm against *P. aeruginosa* and *B. subtilis* followed by 13 mm against *E. coli* and 12 mm against *K. pneumoniae.* Several workers have reported that many plants possess antimicrobial properties including the parts which include flower, bark, stem, leaf, etc. Recently, a number of plants have been reported for antimicrobial properties across the world (Aswar *et al.,* 2007; Koshi Philip *et al.,* 2009, Samy *et al.,* 2000). In the present study also, the methanolic extract of *P. emblica* showed the antibacterial activity against the fish pathogens. Sankannavar and Patil (2012) studied the *in vitro* antibacterial activities of methanolic and ethanolic extracts of *P. emblica* against four food borne pathogens viz., *Staphylococcus aureus, Salmonella typhi, Vibrio cholerae* and *P. aeruginosa* and found with varied degree of zone of inhibition. Similar to the previous observations, in the present study also *P. emblica* methanolic extracts displayed diverse spectrum of activity against *P. aeruginosa.* In addition to this, we studied the bio-efficacy against *K.*

pneumoniae, B. subtilis and *E. coli* and obtained varied zone of inhibition against all the tested pathogens.

Table 2: Antibacterial efficacy of methanolic extract of *P. emblica* L.

S.No	Sample Concentration (μg/ml)	Std.30 (μg/ml)	Zone of inhibition (mm)				
			Solvent	*E. coli*	*P. aeruginosa*	*K. pneumoniae*	*B. subtilis*
1.	25	6	00	5	5	8	7
2.	50	8	00	7	8	9	10
3.	75	9	00	10	9	11	13
4.	100	10	00	13	15	12	15

Standard – Chloramphenicol

Raghu and Ravindra (2010) studied the antibacterial potential of *P. emblica* with 5, 10 and 20 mg/ml diethylether, cloroform and methanolic extracts against *S. aureus* and *K. pneumoniae*. They observed the maximum zone of inhibition (28 mm) against *S. aureus* and 26 mm against *K. pneumoniae*) at 20 mg/ml of *P.emblica* methnaolic extracts. But in the present study we observed 12 mm zone of inhibition against *K. pneumoniae* at 100 μg/ml of *P. emblica* methanolic extracts. Saeed and Tariq (2007) reported that *P. emblica* posse's potent antibacterial activity against *Escherichia coli, K. ozaenae, Klebsiella pneumoniae, Proteus mirabilis, Pseudomonas aeruginosa, S. paratyphi* A, *S. paratyphi* B and *Serratia marcescens*. In addition to previous observation the present supplements the antibacterial potential of *P. emblica* by inhibiting the growth of potent fish pathogens viz., *Klebsiella pneumoniae, Bacillus subtilis, Escherichia coli* and *Pseudomonas aeruginosa*.

4. CONCLUSION

The results of present study suggest that the methanolic extracts of *P. emblica* possesses significant antibacterial activities and confirms the traditional use of this plant in the treatment against the infectious diseases. The structural property of the compounds present in *P. emblica* and their further pharmacological property can be analysed in the future work which in turn will be used as the basic study for the production of novel ayurvedic and allopathic medicines from *P. emblica.*

REFERENCES

Akerele, O., 1993. Summary of WHO Guidelines for the Assessment of Herbal Medicines *Herbal Gram* 22: 13-28.

Alarcon-Aguilar, mag F.J., Roman-Ramos. R., J.L. Flores-Saenz, Aguirre-Garcia, F., 2002. Investigation on the Hypoglycaemic Effects of Extracts of Four Mexican Medicinal Plants in Normal and Alloxan-diabetic Mice. *Phytotherapy Research* 16: 383-386.

Aswar, P.B., Khadabad S.S., Kuchekar,, B.S., Rajurkar, R.M., Saboo, S.S., Javarkar, R.D., *In Vitro* Evaluation of Anti-bacterial and Antifungal activity of Vitex negundo (Verbenaceae), *Phytotherapy Research* 2007; 17: 129-139.

Burkill, I.H., 1966. A Dictionary of the Economic Products of the Malay Peninsula, Ministry of Agriculture and Co-operatives, Kuala Lumpur Vol. 1.

Calixto, J.B., Santos, A.R.S., Filho, V.C., Tunes R.A., 1998. A review of the plants of the genus *Phyllanthus*: Their chemistry, pharmacology, and therapeutic potential. *J. Med. Biol. Res.*31: 225–258.

Dhale, D.A., and Magle, U.P., 2011. Phytochemical Screening and Antibacterial Activity of *Phyllanthus emblica* (L.). *Science Research Reporter* 1(3): 138-142.

Harborne, J.B., 1984. Phytochemical methods, Chapman Hall, London, pp 100-101.

Hardi, A.L., Uddin, M.I., 2004. Seasonal variation in the intestinal bacterial flora of hybrid Tilapia (Oreochromis niloticus x Oreochromis aureus) culture in earthen pond in Saudi Arabia. *Aquaculture* 229 (124), 37- 44.

Khopde, S.M., Priyadarshini, K.I., Guha, S.N., Satav, J.G., Venkatesan, P., Rao, M.N.A., 2000. Biosci. *Biotechnol. Biochem* 64, 503-509.

Koshi Philip, Sri Nurestri Abd Malek, Wirakarnain Sani, Sim Kae Shin, Saravana kumar. Antimicrobial Activity of medicinal plant from Malaysia. *American Journal of Applied Science* 2009; 6 (8): 1613-1617.

Kusirisin, W., Srichairatanakool, S., P.Lerttrakarnnon, 2009. Antioxidative activity, polyphenolic content and anti-glycation effect of some Thai medicinal plants traditionally used in diabetic patients. *Medicinal Chemistry* 5(2):139-47.

Lakshmi Narasimhudu, C., Venkata Raju, R.R., 2012, Phytochemical constituents of Phllanthus species (Euphorbiaceae) From Eastern Ghats of Andhra Pradesh, India, IRJP,ISSN 2230-8407, 3(5):184-200.

Mukherjee, K.L., 2004. Medical Laboratory Technology. Tata McGraw Hill Publishing Company Ltd, New Delhi.

Perry, L.M., 1980. Medicinal Plants of East and Southeast Asia: Attributed Properties and Uses, MIT Press, Cambridge. pp. 149-150.

Piva, R., Penolazzi, L., Borgatti, M., 2009. Apoptosis of human primary osteoclasts treated with molecules targeting nuclear factor-kappaB. Annals of the New York Academy of Sciences 1171:448-456.

Raghu HS and Ravindra P. Antimicrobial activity and phytochemical study of *Phyllanthus emblica* Linn. *International Journal of Pharmaceutical Studies and Research* 1(1): 30-33.

Rastogi, R.P., Mehrotra, B.N., 2002. Glossary of Indian Medicinal Plants. National Institute of Science Communication, New Delhi, India.

Samy RP, Ignacimuthu S. Antibacterial activity of some folklore medicinal plants used by tribals in Western Ghats in *India Journal of Ethnopharmacology* 2000; 69: 63-71.

Sankannavar, S.H., Patil, C.G., 2012. *In Vitro* Studies on Diversity of Antibacterial Activity in Some Species of Phyllanthus for Human Pathogenic Bacteria. *Asian J. Exp. Biol. Sci.* 3 (3).

Saeed, S. and P. Tariq, 2007. Antibacterial activities of *Emblica officinalis* and *Coriandrum sativum* against Gram negative urinary pathogens. Pak J 10(6-7): 583-9. *Pharm Sci.*, 20(1): 32-5.

Summanen, J.O., 1999. A Chemical and Ethnopharmalogical Study on *Phyllanthus emblica* (Euphorbiaceae), Division of Pharmacognosy, Department of Pharmacy, University of Helsinki, Finland.

Thakur, R.S., Puri, H.S., Husain, Akhtar, 1994. Major Medicinal Plants of India Central Institute of Medicinal and Aromatic Plants, *Phytochem Anal.* 18(3):251-255.

Treadway, and Linda, 1994. Amla Traditional food and medicine. HerbalGram. *The Journal of the American Botanical Council* 31: 26.

5

Bioactive Antimalarial Agents from Plants

Francis Stanly Pradeep, Vijayakumar Usha Rani and Bhathini Vaikuntavasan Pradeep

1. INTRODUCTION

Malaria remains one of the most serious tropical diseases and kills an estimated one to two million people each year worldwide (Snow *et al.*, 2005). Spread of drug resistance has played an important role in the occurrence and severity of epidemic diseases in the world (Majori, 2004). As a consequence of resistance to drugs such as quinine, chloroquine, primaquine and mefloquine, the vehemence of malaria has increased in many endemic regions of the world (Schlitzer, 2007; Basco *et al.*, 1994). One of the biggest challenges to malaria control is the emergence and spread of drug resistant *P. falciparum* strains. Although human malaria transmitted by female Anopheles mosquitoes has four Plasmodium species as its aetiological agents *P. falciparum*, *P. vivax*, *P. ovale*and *P. malariae*, the most widespread and severe disease is caused by *P. falciparum*, which transiently infects the liver before invading red blood cells of the mammalian host. Clinical manifestations occur at the erythrocytic stage and can include fever, chills, prostration and anaemia, as well as delirium, metabolic acidosis, cerebral malaria and multi-organ system failure, which may be followed by coma and death (Fidock *et al.*, 2004; Deprez-Poulain *et al.*, 2005, De Sa *et al.*, 2009).The success of the natural product-based antimalarials such as quinine and artemisinin has showed the importance of plants as potential source of antimalarial agents. Consequently, numerous efforts to identify antiplasmodial activities from plants traditionally used as remedies for fever and malaria (Chen *et al.*, 1997; Havsteen, 2002; Lacroix *et* *al*., 2011), coupled with bioassay guided isolations of active principles have resulted in the identification of limonoids, coumarines, alkaloids, flavanoids (including chalcones), quassinoids, xanthones, terpenes, quinones and peptides as antimalarial principles (Kaur *et al*., 2009).

Indeed, the vast majority of the existing antimalarial chemotherapeutic agents are based on natural products, and this fact anticipates that new leads may certainly emerge from the tropical plant sources, since biological chemodiversity continues to be an important source of molecular templates in the search for antimalarial drugs (Portet, 2007). Validation of traditional plant remedies presents limitations such as prioritization of plant species for research, lack of informations on ethnobotany of these plants (location and abundance, parts used,form of use, duration of treatment), definition of dosages due to variation on the concentrations of active ingredients in a plant species (Willcox *et al*., 2005). It should also be stressed that a basic requirement for the validation of a medicinal plant is the standardization of plant extracts to be evaluated, including the identification and quantification of chemical and/or biological markers to assure the development of efficient and safe phytomedicines in a short time and at low cost (Ronan *et al*., 2009; Harborne *et al*., 2001).

2. ANTIPLASMODIAL ACTIVITY OF TEN PLANT BIOACTIVE COMPOUNDS

2.1. Dimericphenylanthraquinones

The medicinal implications of the *Bulbine* species have been well known for centuries and were used as early as the eighteenth century. The British and Dutch settlers of Southern Africa utilized the species for various medicinal aliments (Coopoosamy *et al*., 2000). From the roots of the African plant *Bulbine frutescens* (Asphodelaceae), two unprecedented novel dimericphenylanthraquinones, named joziknipholones A and B, possessing axial and centrochirality, were isolated, together with six known compounds (Abegaz *et al*., 2002).Based on the recently revised absolute axial configuration of the phenylanthraquinones, knipholone and knipholone anthrone, the new dimers were attributed to possess the P-configuration (i.e., with the acetyl portions below the anthraquinone plane) at both axes in the case of Joziknipholone A, whereas in Joziknipholone B, the knipholone part was found to be M-configured. Joziknipholones A and B are active against the chloroquine resistant strain K1 of the malaria pathogen, *Plasmodium falciparum*, and show moderate activity against murine leukemic lymphoma L5178y cells(Bringmann *et al*. 2008).

2.2. Sesquiterpenes

Two of these (17%) were found to have IC50<5 μg/ mL for either CQ sensitive (NF54) or resistant (ENT30) strains andthese were the methanol and water root

Table: 1. Antiplasmodial activity of constituents from ten medicinal plants.

Family	Species	Extract	Antiplasmodial Activity (IC50, mg/mL)	Compound	Reference
Asphodelaceae	*Bulbine frutescens*	MeOH (aerial parts)	0.14±0.2	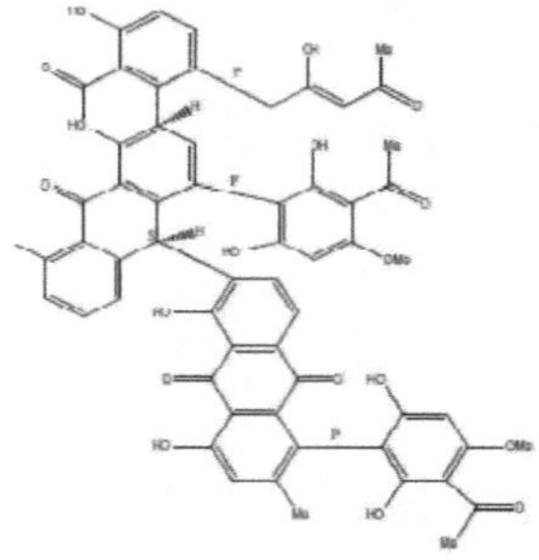 Phenylanthraquinones: Joziknipholone A	Bringmann *et al.*, 2008
Cyperaceae	*Cyperus articulatus L.*	MeOH (bark)	4.84±0.56	Sesquiterpenes: Mustakone 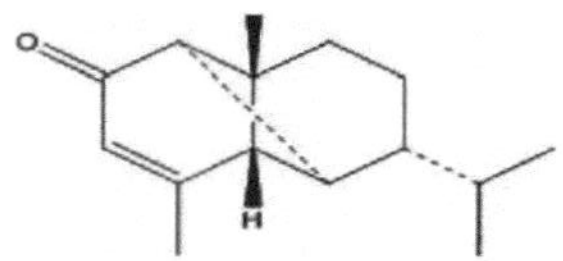 Sesquiterpenes: Mustakone	Rukunga *et al.*, 2008
Cyperaceae	*Scleria striatinux*	CH_2Cl_2; MeOH extract	80.4±0.8	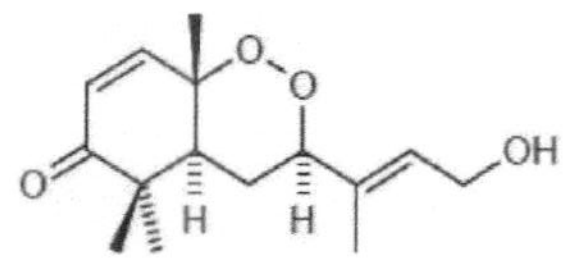 Okundoperoxide	Erfange *et al.*, 2009

Family	Species	Extract	Antiplasmodial Activity (IC50, mg/mL)	Compound	Reference
Fabaceae	*Erythrina burttii*	Acetone stem extract	19.4±1.8	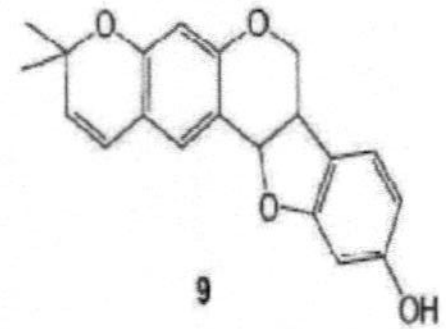 9 Pterocarpan-Calopocarpin	Yenesew *et al*., 2012
Fabaceae	*Bowdichia nitida*	MeOH(aerial parts)	1 μM	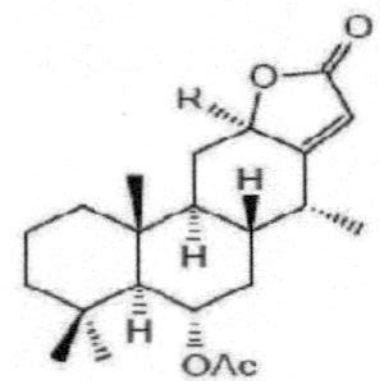 Sucutiniranes A and B	Matsuno *et al*., 2008
Gentianaceae	*Fagrae afragrans*	MeOH (aerial parts)	116.6 ± 9.4	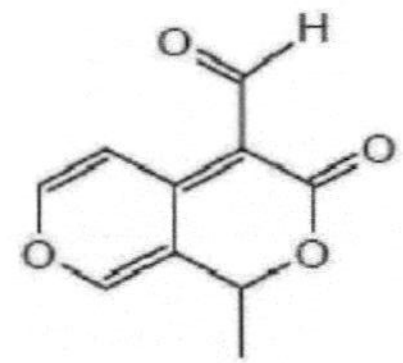 Fagraldehyde	Jonville *et al*., 2008

Family	Species	Extract	Antiplasmodia lActivity (IC50, mg/mL)	Compound	Reference
Meliaceae	*Ekebergia capensis*	MeOH (aerial parts)	6.0±0.5	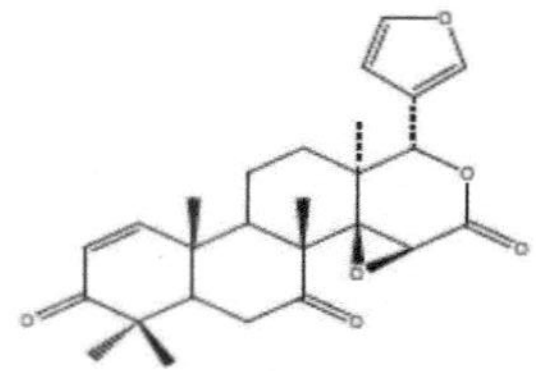 Triterpenoidlimonoid-deacetoxy-7-oxogedunin	Murata *et al.*,2008
Mimosaceae	*Albizia gummifera*	$CHCl_3$ (stembark)	0.16±0.05	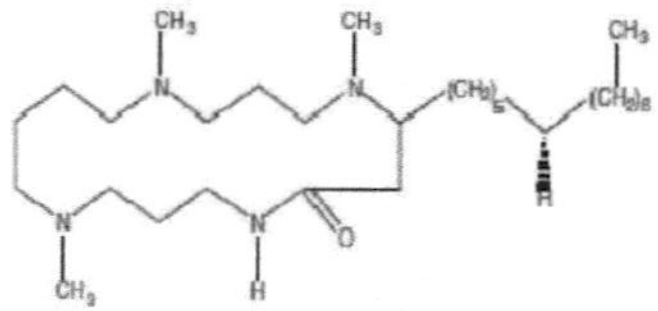 Sperminealkaloids	Rukunga *et al.*, 2007
Lamiaceae	*Salvia radula*	CH_2Cl_2; MeOH (aerialparts)	3.91±0.52	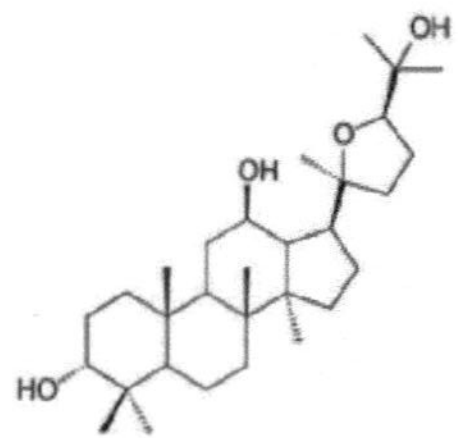 Beulafolientriol Oxide	Kamatou *et al.*, 2006

Family	Species	Extract	AntiplasmodialActivity (IC50, mg/mL)	Compound	Reference
				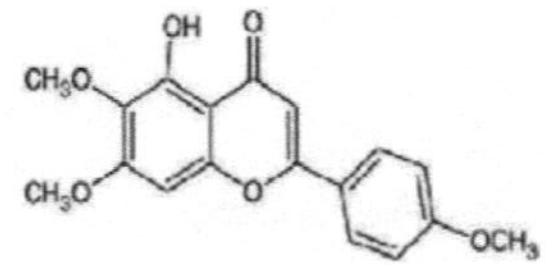 Salvigenin	
Zingiberaceae	*Siphonochilus aethiopicus*	EtOAc (rhizomes)	2.90±0.28	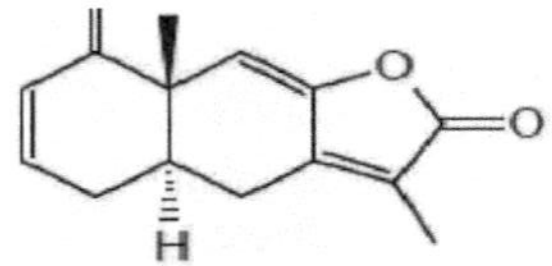 Furanoterpenoids	Lategan *et al.*, 2009

bark extract of *Zanthoxylum chalybeum* Engl. (Rutaceae) and the methanol rhizome extract of *Cyperus articulatus L.* (Cyperaceae).Two sesquiterpenes, corymbolone and mustakone, isolated from the chloroform extract of the rhizomes of *Cyperus articulatus*, exhibited significant anti-plasmodial properties(Rukunga *et al.*, 2008).

2.3. Okundoperoxide

The CH_2Cl_2-MeOH extract of the roots of *Scleriastriatinux*de Wild (syn. *S. striatonux*)(Cyperaceae), a local spice in parts of Cameroon, was shown to exhibit low activity against both D_6chloroquine-sensitive (IC50= 80.4 µg/mL) and W_2 resistant (IC50 = 89.4 µg/mL) strains of *P.falciparum.* A new and skeletally unique bicyclofarnesyl sesquiterpene endoperoxide, named okundoperoxide was isolated by bioassay-guided fractionation of extracts from this plant (Erfange *et al.*, 2009).

2.4. Pterocarpan-Calopocarpin

Erythrina species contain prenylated flavonoids with anti-infective activities (Mitscher *et al.*, 1987).Yenesew *et al.*, (2003, 2004)reported that several flavonoids with antiplasmodial activity from the roots and stem bark of *Erythrina abyssinica* and the stem bark of *Erythrina sacleuxii* and the stem bark of *Erythrina sacleuxii*(Andayi *et al.*, 2006).Among the five *Erythrina* species present in Kenya, *E. abyssinica* is the most widely used in traditional medicine where it has been used for treatment of malaria and microbial infections (Kokwaro, 1993).Flavonoids have been isolated and characterized from many medicinal plants used to treat infections around the world (Nundkumar and Ojewole, 2002), among which are *Erythrina species.* Since the identification of the chalconelicochalcone A, as a lead structure in antimalarial drug development (Ziegler *et al.*, 2004), there is a growing interest in antimalarial activity of chalcones and other flavonoids.The acetone extract of the root bark of *Erythrina burttii* showed *in vitro* antiplasmodial activity against the chloroquine-sensitive (D_6) and chloroquine-resistant (W_2) strains of *Plasmodium falciparum* with IC50 values of 0.97±0.2 and 1.73±0.5 µg/mL respectively (Yenesew *et al.* 2012).

2.5. Sucutiniranes A and B

Japanese researchers have isolated two new cassane-type diterpenes, sucutiniranes A and B from the seeds of *Bowdichia nitida*(Fabaceae), a Brazilian Amazon species commonly known as "sucupira". Compound showed promising *in vitro* antiplasmodial activity against parasite *P. falciparum* 3D7 (IC50 = 1µM) and a good selectivity index with regard to the cytotoxicity on COLO201 cells (IC50 > 250 µM), whereas other compounds were not active at a concentration of 3µM (chloroquine: IC50 = 0.019 µM) (Matsuno *et al.*, 2008).

2.6. Fagraldehyde

Fagraea fragrans is a large evergreen tree in the family Gentianaceae. It is native to South-east Asia. Its trunk is dark brown, with deeply fissured bark, looking somewhat like a bittergourd. Its yellowish flowers have a distinct fragrance and the fruits of the tree are bitter tasting red berries, which are eaten by birds and fruit bats. A secoiridoida glycone with an atypical skeleton, named fagraldehyde, were isolated from the bark and leaves of *Fagraea fragrans*collected in Cambodia. Compound was weakly active *invitro* against *P. falciparum*, exhibiting an IC50 value of 116.6 ± 9.4 ìM (W2 strain) (Jonville *et al.*, 2008).

2.7. Ekeberins A

Ekebergia capensis is a tree in the Meliaceae family. It is commonly known as the Cape Ash. Its range extends from the Eastern Cape of South Africa to Sudan and Ethiopia. From the stem bark of *Ekebergia capensis*, 10 new triterpenoid compounds, ekeberins A were isolated together with 17 known compounds. Several of these compounds were screened in vitro against both chloroquine (CQ)-sensitive and resistant *Plasmodium falciparum* isolates and were found to exhibit moderate antiplasmodial activity, with compounds-20 (7-deacetoxy-7-oxogedunin) and 27 (2-hydroxymethyl-2, 3, 22, 23-tetrahydroxy-2, 6, 10, 15, 19, 23-hexamethyl-6, 10, 14, 18 tetracosatetraene) showing IC50 values of 6 and 7 mM, respectively(Murata *et al.*, 2008). Muregi*et al.* (2004) found similar *in vitro* antiplasmodial activity for *E. capensis* collected fom Kisii district in Kenya. A limonoid compound (4), (Oxogedunin) (Bevan *et al.*, 1963) and a triterpenoid (5),(2-hydroxymethyl-2,3,22,23-tetrahydroxy-2,6,-10,15,19,23-hexamethyl-6,10,14,18 tetracosatetraene) (Nishiyama *et al.*, 1996, 1999) both isolated from the plant showed potent IC50 activities as low as 6 and 7 lM, respectively.

2.8.Spermine alkaloids

The discovery of quinine from *Cinchona species* and artemisinin from *Artemisia annua* has stimulated interest in medicinal plants as sources of new antimalarial drugs. The stem bark of *A. gummifera* is used in Kenya by traditional healers for preparation of herbal remedy for malaria.One of them, *A. gummifera* methanol stem bark extract was found to have IC50<5 µg/ml for all the isolates. All other fractions apart from the alkaloidal fraction showed low activity with IC50 above 3 µg/mL. The alkaloidal fraction exhibited strong activity against NF54 and ENT30 with IC50 of 0.16±0.05 and 0.99±0.06µg/mL, respectively. Five known spermine alkaloids were isolated from the alkaloidalfraction. These alkaloids exhibited activities against NF54 and ENT30 with IC50 ranging from 0.09±0.02 to 0.91±0.10 µg/mL. Fourof the alkaloids were further evaluated for in vivo activity against rodent malaria parasite *Plasmodium berghei*(Rukunga *et al.*, 2007; Oliveira *et al.*, 2009).

2.9.Salvigenin

Salvia radula is an herbaceous perennial native to South Africa, growing at elevations from 1,300 to 1,900 m (4,300 to 6,200 ft). As *Salvia radula* (Lamiaceae) exhibited the best antiplasmodial activity (IC50 = 3.91 μg/mL) among 17 *Salvia* species used in traditional medicine in South Africa, it was selected for the bioassay-guided isolation of the active compounds. Salvia species have been used in traditional medicines in China, South Africa and many other countries against various infectious and inflammatory diseases and to treat malaria, hard swellings, abscesses, calluses, warts or cancer (Watt and Breyer- Brandwijk, 1962; Ulubelen *et al.*, 1999; Shoemaker *et al.*,2005).Fractionation of a 1:1 methanol/chloroform extract of this species afforded betulafolientriol oxide and salvigenin, which displayed similar or lower antiplasmodial activity [IC_{50} values = 4.95 and 24.60 μg/mL (10.4 and 74.9 μM), respectively] when compared to the crude solvent extract(Kamatou *et al.*, 2006).

2.10.Furanoterpenoids

Wild ginger (*Siphonochilus aethiopicus*(Schweinf) B.L Burtt)is an herb of the Zingiberaceae family growing in the middle belt region of Nigeria. Its perennial tuberous roots give rise to annual leafy shoots which grow to about 35 cm high after flowering and are common throughout the West African region and elsewhere in tropical Africa (Burkill, 2000).

From the crude plant extract of the medicinal plant *Siphonochilus aethiopicus* which showed a good *in vitro* activity against strains of *P. falciparum* with IC_{50} values of 2.9 and 1.4μg/mL,respectively, a bioassay-guided fractionation led to the isolation of furanoterpenoids with good *invitro* antiplasmodial activity(Lategan *et al.*, 2009; Carmen *et al.*, 2009).

Perspectives on the Development of New Drugs and Phytomedicines for Malaria

The antiplasmodial natural products focused in this review can be divided into two groups: one of highly active compounds, of complex structures, for which no possibility of practical synthesis can be forseen, and another one, of moderate to low activity but of relative simple structures and, therefore, their synthesis and/or of their analogues could be undertaken. In many parts of the world the parasites have developed resistance to a number of antimalarials such as chloroquine and derivatives, the most widely used treatment for malaria, and so there is an urgent need to discover new compounds with an original mode of action. Ideally, plant extracts and compounds effective at the blood stage of the malaria parasite should have strong *in vitro* antiplasmodial and *in vivo* antimalarial activities with good selectivity for the malaria parasites in cytotoxic assays. Dependence on medicinal plants as the source of medicine is prevalent in developing countries where traditional medicine plays a major role in health care. Plant species producing bioactive compounds of the first group are potential candidates for the development of phytomedicines, while the second group would represent templates to synthetic drugs.

3. CONCLUSION

Considering the many antiplasmodial activities observed for all potential plant extracts, it can be stated that in fact there are optimistic perspectives on the continuing investigation of plants used in traditional medicines for the treatment of malaria, and they will certainly lead the scientific community to the discovery of more new efficient molecular templates and phytomedicines for this disease. Ethnopharmacological approach appears to be a promising way to find plant secondary metabolites that could be used as templates for designing new derivatives with improved properties.

ACKNOWLEDGEMENT

The authors are sincerely grateful to the management, Karpagam University, Coimbatore, Tamil Nadu, India – 641 021 for encouragement and support.

REFERENCES

Abegaz, B.M., Bezabih, M., Msuta, T., Brun, R., Menche, D., Muhlbacher, J.,Bringmann, G., 2002.Gaboroquinones A and B and 40-O-demethylknipholone-40-O-b-D-glucopyranoside, phenylanthraquinones from the roots of *Bulbinefrutescens.Journal of Natural Products* 65.1117–1121.

Andayi, A.W., Yenesew, A., Derese, S., Midiwo, J.O., Gitu, P.M., Jondiko, O.J., Akala, H., Liyala, P., Wangui, J., Waters, N.C., Heydenreich, M., Peter, M.G., 2006. Antiplasmodial flavonoids from *Erythrinasacleuxii. PlantaMedica* 72: 187–189.

Basco, L.K., Mitaku, S., Skaitsounis, A., Ravelomanantsoa, N., Tillenquin, F., Koch M. 1994. *Invitro* activities of fluroquinoline and acridone alkaloids against *Plasmodium falciparum.Antimicrobial Agents and Chemotheraphy*.38: 1169-71.

Bavovada, R., Blaskom, G., Shienm, H.L., Pezzutom, J.M., Cordell, G.A., 1990. Spectralassignment and cytotoxicity of 22-hydroxytingenone from *Glyptopetalums clerocarpum. PlantaMedica*. 56: 380–382.

Bringmann, G., Mutanyatta-Comar, J., Maksimenka, K., Wanjohi, J.M., Heydenreich, M., Brun, R., Muller, W.E., Peter, M.G., Midiwo, J.O., Yenesew, A. 2008. Jozikniphololones A and B: the first dimericphenylanthraquinones, from the roots of *Bulbinefrutescens*. Chemistry 14(5), 1420–9.

Burkill, H.M., 2000. Useful Plants of West Tropical Africa.Vol 5.Families S–Z. Crown Agents for Overseas Governments, London.Vol 5. 2nd Edition, pp. 320-321.

Carmen, A., Lategan, William, E., Campbell, Tracy, S., Peter, J. 2009.The bioactivity of novel furanoterpenoids isolated from *Siphonochilusaethiopicus. Journal of Ethnopharmacology*. 121: 92–97

Chen, M., Brogger, C.S., Zhai, L., Rasmussen, M.H., Theander, T.G., Frokjaer, S.,Steffansen, B., Davidsen, J., Kharazmi, A. 1997. The novel oxygenated chalcone, 2,4 dimethoxy- 4-butoxychalcone, exhibits potent activity against humanmalaria parasite *Plasmodium falciparuminvitro* and rodent parasites *Plasmodium berghei* and *Plasmodium yoelii*. *Journal of Infectious Diseases* 176,1327–1333.

Coopoosamy, M., Magwa, M.L., Mayekiso B., 2000. Proceedings: Science 224 Afrian Journal Biotechnology and Society 2000, University of Fort Hare, Bhisho, Eastern Cape,South Africa.

De Sa, M.S., Costa, J.F., Krettli, A.U., Zalis, M.G., Maia, G.L., Sette, I.M., Camara, C.D.,Filho, J.M., Giulietti-Harley, A.M., Ribeiro Dos Santos, R.,Soares, M.B.,2009.Antimalarial activity ofbetulinic acid and derivatives in vitro against *Plasmodium falciparum*and *invivo*in *P. berghei*infectedmice. *Parasitologicial Research*. 105: 275-279.

Deprez-Poulain, R., Melnyk, P., 2005. 1, 4-Bis (3-aminopropyl) piperazine libraries: from the discovery of classical chloroquine-like antimalarials to the identification of new targets. *Combinatorial Chemistry and High Throughput Screening*. 8: 39–48.

Efange, S.M., Brun, R., Wittlin, S., Connolly, J.D., Hoye, T.R., McAkam, T., Makolo, F.L., Mbah, J.A., Nelson, D.P., Nyongbela, K.D., Wirmum, C.K. 2009.Okundoperoxide, a bicyclic cyclofarnesylsesquiterpeneendoperoxide from *Scleriastriatinux* with antiplasmodial activity.*Journal of Natural Products*.72: 280–283.

Fidock, D.A., Rosenthal, P.J., Croft, S.L.,Brun, R.,Nwaka, S. 2004. Antimalarial drug discovery: efficacy models for compound screening. Nature Reviews Drug Discovory.3: 509–520.

Harborne, J.B. Twenty-five years of chemical ecology. 2001. *Natural Products Reports* 18, 361-379.

Havsteen, B.H., 2002. The biochemistry and medical significance of the flavonoids.*Pharmacology & Therapeutics*. 96: 67–202.

Jonville, M.C., Capel, M., Frederich, M., Angenot, Luc., Dive, G., Faure, R., Azas, N.,Ollivier,E.2008.Fagraldehyde, a secoiridoid isolated from *Fagraeafragrans*. Journal of Natural Products. 71: 2038–2040.

Kamatou, G.P.P., Van Zyl, R.L.,Davids, H.,Heerden, F.R.V.,Lourens, A.C.U.,Viljoen, A.M. 2008.Antimalarial and anticancer activities of selected South African Salvia species and isolated compounds from *S. radula*. *South African Journal of Botany*.74: 238–243.

Kaur, K., Jain, M., Kaur, T., Jain, R., 2009. *Antimalarials* from nature. *Bioorganic and Medicinal Chemistry Letters*. 17: 3229–3256.

Kokwaro, J.O., 1993. Medicinal plants of East Africa. Kenya Literature Bureau, Nairobi.

Lacroix, D., Prado, S., Kamoga, D., Kasenene, J., Namukobe, J., Krief, S., Dumontet, V.,Mouray, E., Bodo, B., Brunois, F., 2011. Antiplasmodial and cytotoxic activities of medicinal plants traditionally used in the village of Kiohima, Uganda. Journal of Ethnopharmacology. 133: 850–855.

Lategan, C.A., Campbell, W.E., Seaman, T., Smith, P.J.,2009. The bioactivity of novel furanoterpenoids isolated from *Siphonochilusaethiopicus*. *Journal of Ethnopharmacology*.121: 92–97.

Majori, G. 2004. Combined antimalarial therapy usingartemisinin. *Parassitologia*. 46: 85–87.

Matsuno, Y., Deguchi, J., Hirasawa, Y., Ohyama, K., Toyoda, H., Hirobe, C., Ekasari, W., Widyawaruyanti, A., Zaini, N.C., Morita, H.2008.Sucutiniranes A and B, new cassane-type diterpenes from *Bowdichianitida*.Bioorganic and Medicinal Chemistry *Letters*.18: 3774–3777.

Mitscher, L.A., Drake, S., Gollapudi, S.R., Okwute, S.K., 1987. A modern look at folkloricuse of anti-infective agents. *Journal of Natural Products* 50, 1025–1040.

Murata, T., Miyase, T., Muregi, F.W., Naoshima-Ishibashi, Y., Umehara, K., Warashina, T., Kanou, S., Mkoji, G.M., Terada, M., Ishih, A., 2008.Antiplasmodialtriterpenoids from *Ekebergiacapensis*. *Journal of Natural Products* 71, 167–174.

Muregi, F.W., Chhabra, S.C., Njagi, E.N., Langat-Thoruwa, C.C., Njue, W.M., Orago, A.S., Omar, S.A., Ndiege, I.O., 2004. Anti-plasmodial activity of some Kenyanmedicinal plant extracts singly and in combination with chloroquine. *Phytotherapy Research* 18:379–384.

Nishiyama, Y., Moriyasu, M., Ichimaru, M., Kato, A., Mathenge, S.G., Nganga, J.N., Juma, F.D., 1999.Absolute configurations of two acyclic triterpenoids from *Ekebergiacapensis. Phytochemistry.* 52: 1593–1596.

Nishiyama, Y., Moriyasu, M., Ichimaru, M., Tachibana, Y., Kato, A., Mathenge, S.G.,Nganga, J.N., Juma, F.D., 1996.Acyclic triterpenoids from *Ekebergiacapensis.Phytochemistry.* 42: 803–807.

Nundkumar, N., Ojewole, J.A., 2002. Studies on the antiplasmodial properties of some South African medicinal plants used as antimalarial remedies in Zulu folk medicine. Methods and Findings in *Experimental and Clinical Pharmacology.* 24: 397–401.

Oliveira, A.B., Dolabela, M.F., Braga, F.C., Jacome, R.L.R.P., Varotti, F.P., Povoa, M.M. Plant derived antimalarial agents: new leads and efficient phytomedicines. Part I. Alkaloids. Annais da Academia Brasileria de ciencias. 2009, *in press.*

Portet, B., Fabre, N.,Roumy, V., Gornitzka, H., Bourdy, G., Chevalley, S., Sauvain, M., Valentin, A., Moulis, C. 2007.Activity-guided isolation of antiplasmodialdihydrochalcones andflavanones from *Piper hostmannianum*var. berbicense.*Phytochemistry.*68: 1312–1320.

Ronan, B., Ademir, D., Alaide, B. 2009. Plant-Derived Antimalarial Agents: New Leads and Efficient Phytomedicines. Part II. Non-Alkaloidal *Natural Products Molecules.* 14: 3037-3072.

Rukunga, G.M.,Muregi, F.W., Omar, S.A.,Gathirwa, J.W., Muthaura,C.N., Peter, M.G., Heydenreich, M., Mungai, G.M.2008. Anti-plasmodial activity of the extracts and two sesquiterpenes from *Cyperusarticulatus. Fitoterapia.* 79:188–190

Rukunga, G.M., Muregi, F.W., Tolo, F.M., Omar, S.A., Mwitari, P., Muthaura, C.N., Omlin, F., Lwande, W., Hassanali, A., Githure, J., Iraqi, F.W., Mungai, G.M., Kraus, W., Kofi Tsekpo, W.M.,2007. The antiplasmodial activity of spermine alkaloids isolated from *Albiziagummifera. Fitoterapia.* 78:455–459

Schlitzer, M. 2007. Malaria chemotherapeutics part I: history of antimalarial drug development, currently used therapeutics and drugs in clinical development. *Chemistry and Drug Discovery.* 7: 944–986.

Shoemaker, M., Hamilton, B., Dairkee, S.H., Cohen, I., Campbell, M.J., 2005.Invitro anticancer activity of twelve Chinese medicinal herbs.*Phytotherapy Research.* 19: 649–651.

Snow, R. W., Guerra, C. A., Noor, A. M., Myint, H. Y., Hay, S. I.,2005. The global distribution of clinical episodes of *Plasmodium falciparum* malaria.*Nature.* 434: 214.

Ulubelen, A., Topcu, G., Chai, H.B., Pezzuto, J.M., 1999. Cytotoxicity activityof diterpenoids isolated from *Salvia hypargeia. Pharmaceutical Biology.* 37:148–151.

Watt, J.M., Breyer-Brandwijk, M.G. 1962.The Medicinal and Poisonous Plantsof Southern and Eastern Africa, 2nd edition. E & S Livingstone, Edinburgh.

Willcox, M.L.,Bodeker, G. 2004.Traditional herbal medicines for malaria.*British Medical Journal.* 329: 1156–1159.

Yenesew, A., Hoseah, M., Akala, C., Hannington, T., Carolyne, C., Beatrice, N., Irungue, F., Eyase, L., Maud, K., Bernard, T., Kiremire, J., Johnson, D., Norman, C. 2012.The antiplasmodial and radical scavenging activities of flavonoids of *Erythrinaburttii. ActaTropica.* 123:123-127

Yenesew, A., Induli, M., Derese, S., Midiwo, J.O., Heydenreich, M., Peter, M.G., Akala,H., Wangui, J., Liyala, P., Waters, N.C. 2004. Anti-plasmodial flavonoids from thestem bark of *Erythrinaabyssinica. Phytochemistry.* 65: 3029-3032.

Yenesew, A., Midiwo, J.O., Guchu, S.M., Heydenreich, M., Peter, M.G., 2002. Threeisoflav-3-enes and a 2-arylbenzofuran from the root bark of *Erythrinaburttii.Phytochemistry.* 59:337–341.

Ziegler, H.L., Hansen, H.S., Staerk, D., Christensen, S.B., Hagerstr, H., Jaroszewski, J.W., 2004. The antiparasitic compound licochalcone A is a potent echinocytogenicagent that modifies the erythrocyte membrane in the concentration range where antiplasmodial activity is observed. *Antimicrobial Agents and Chemotherapy.* 48:4067–4071.

6

Evaluation of Antibacterial Activity of *Phyllanthus reticulatus* Poir

Renisheya Joy Jeba Malar, T., Johnson, M., Siva Raman, A. and Arul Starlin, I.

1. Introduction

India throughout its long history has accumulated a prosperous practical acquaintance of the use of medicinal plants for the treatment of various diseases. Chemical studies on Indian medicinal plants make available a priceless material foundation for the discovery and development of new drugs of natural origin. Contrary to the synthetic drugs, antimicrobials of plant origin are not associated with many side effects and have an enormous therapeutic potential to cure many infectious diseases (Bibitha *et al.*, 2002; Maghrani *et al.*, 2005; Doughari, 2006). Plants are used as a medicine in different countries and are a source of many potent and powerful drugs (Kambizi and Afolayan, 2008). Extraction and characterization of several active compounds from these green factories have given birth to some high activity profile drugs (Mandel *et al.*, 2007). *Phyllanthus* is a medicinally important genus belongs to the family Euphorbiaceae. A number of plants of this family are observed to be beneficial for the treatment of liver ailments. *Phyllanthus reticulatus* Poir. is a large climbing shrub growing from 8-10 feet in height. It is used in the traditional medicinal system of our country for treatment of a variety of ailments including small pox, syphilis, asthma, diarrhoea, and bleeding from gums (Kirtikar and Basu, 2003; The Wealth of India, 2005). Plants of *Phyllanthus* are reported to contain lignans, flavonoids, triterpenoids, alkaloids, polyphenolic compounds which possess significant activity against hepatitis B virus responsible for hepatotoxicity causing fatal liver diseases (Anjenenlu *et al.*, 1973: Joshi *et al.*, 1991; Yoshida *et al.*, 1982). As a part of our continuing interest in the study of biological activities, the present investigation was focussed on phytoprofile and antibacterial activity of *P. reticulatus* against the selected pathogens (*E. coli, P. aeroginosa, K. pneumoniae, B. subtilis).*

2. MATERIALS AND METHODS

2.1. Collection of Plant Material

The aerial parts of *Phyllanthus reticulatus* Poir. were collected from the natural habitats at Kalakadu, Tirunelveli, Tamil Nadu, India. The plant materials were washed under running tap water to remove the surface contaminants and the leaves and stem were separated mechanically. The separated parts were air dried under shade. The dried sample was powdered using mechanical grinder and used for further extraction.

2.2. Extraction of Plant Material

10 g of air dried powder of the sample was extracted with 60 ml of solvents viz., petroleum ether, chloroform, acetone, methanol and aqueous. The sample was kept with solvents under dark for 72 h with intermittent shaking. After incubation, the solution was filtered through filter paper and the filtrate was collected (crude extracts).

2.3. Preliminary Phytochemical Screening

Preliminary phytochemical screening of various extracts of *P. reticulatus* was carried out to reveal the presence of steroids, terpenoids, cardiac glycosides, saponins, tannins, alkaloids, phenolics and amino acids (Harborne, 1998).

2.4. Antibacterial Activity

Screening of antibacterial activity was performed by disc diffusion method (Mukherjee, 2004). 200 µg/ml of five different extracts (acetone, chloroform, petroleum ether, methanol and aqueous) of *P. reticulatus* were screened for antibacterial studies against selected fish pathogens viz., *E. coli, P. aeruginosa, K. pneumonia* and *B. subtilis* (Hardi *et al.*, 2004). Commercially available antibiotic disc Amikacin was implanted along with the crude extract disc on the surface of the Muller-Hinton agar plates which is used as a positive control. The inoculated plates were incubated at 37°C for 18-24 hours and the zone of inhibition was measured and the results were tabulated.

3. RESULTS AND DISCUSSION

In the present study, preliminary phytochemical screening of eight different metabolites (steroids, saponins, phenols, tannin, alkaloids, anthraquinone, amino acids and terpenoids) were tested in five different extracts. Thus out of 40 (8 x 5 = 40) tests for the presence or absence of the above compounds, only 12 gave positive results and the remaining 28 gave negative results. The 12 positive results show the presence of steroids, saponins, tannin, anthraquinone, amino acids, terpenoids, phenols and alkaloids. Among the five different extracts, methanolic extract of *P. reticulatus* showed the presence of maximum compounds viz., alkaloids, saponins, phenolics and tannins (4/8) whereas chloroform and aqueous extract showed the presence of three compound each. Petroleum ether and acetone extract showed the presence of only one compound steroid (Table 1). Antibacterial activity of *P. reticulatus* was screened in five different extracts (Petroleum ether, chloroform, acetone, methanol and aqueous) against four different pathogens (*E. coli, P. aeruginosa, K. pneumoniae, B.subtilis*) and the results were tabulated in Table 2. Among these 200 μg/ml of acetone extract of *P. reticulatus* showed the maximum zone of inhibition against *E. coli* (15 mm). Chloroform extract of *P. reticulatus* showed moderate inhibitory activity against all the pathogens *P. aeroginosa* (12 mm), *E. coli* (11 mm), *K. pneumoniae* (9 mm), *B. subtilis* (8 mm). The acetone, methanol and aqueous extract of *P. reticulatus* failed to show the inhibition against *B. subtilis*. Petroleum ether and aqueous extract showed no activity against *P. aeruginosa.*

Lakshmi and Raja (2012) revealed the presence of alkaloids, flavonoids, flavones, glycosides, phenols, lignins, anthocyanides and triterpenoids in the methanolic extract of *P. reticulatus* from the Eastern Ghats of Andhra Pradesh. In the present study also, we confirmed the methanolic extracts showed the presence of alkaloids and phenolics in *P. reticulatus* from Western Ghats, Tamil Nadu. But we failed to observe the presence of flavonoids in *P. reticulatus* from Western Ghats, Tamil Nadu. Aqueous extract showed the presence of flavones, alkaloids, phenols, saponins, glycosides and triterpenoids. The present study results were directly coincided with Lakshmi and Venkata Raja's observations. Petroleum ether extract determined the presence of alkaloids, anthocyanins and steroids. The present observations also confirmed the steroids and absence of alkaloids. The results of the present study supplements the previous observation, at the same time it is contrary to the previous observations. The variation in the phytoprofile may be due to the locality/habitat difference. Vaghasiya *et al*., (2011) reported the presence of cardiac glycosides, tannins, saponins, alkaloids and steroids in the methanolic extract of *P. reticulatus.* In comparison to the previous work, methanolic extract of *P. reticulatus* showed the presence of maximum number of secondary metabolites except steroids. Therefore the results indicate that the extracts of *P. reticulatus* might have the possibility of containing components of biological activity. Chloroform and methanolic extract of *P. reticulatus* showed maximum degree of bacterial inhibition activity due to the presence of saponins, phenolics and alkaloids. Saponins are known to have antioxidant, anticancer, anti-inflammatory, antibacterial and anti-fungal properties (De-Lucca *et al.*, 2005; Mandal *et al.*, 2005; Manjunatha, 2006). Alkaloids are used in medicine in the form of salts including anti-arrhythmic, anti-cholinergic, anti-tumor, vasodilating, anti-hypertensive, cough, anesthetic, anti-microbial and anti-protozoal agents (URL, 2012a). Phenolic compounds exhibit a wide range of physiological properties such as anti-inflammatory, anti-microbial, anti-allergenic, antioxidant, cardioprotective and vasodilatory effects (Manach and Mazur, 2005). The present study results confirmed the traditional usage and the previous observations on the biological activities. Further studies are required to confirm the active compounds responsible for bioactivity.

Table 1: Preliminary Phytochemical screening of *P. reticulatus*

Solvents	Pet. ether	Chloroform	Acetone	methanol	Aqueous
Steroids	+	-	+	-	-
Alkaloids	-	-	-	+	-
Saponins	-	+	-	+	-
Anthraquinone	-	-	-	-	-
Aminoacid	-	+	-	-	-
Phenolics	-	+	-	+	+
Tannins	-	-	-	+	+
Flavonoids	-	-	-	-	+

Table 2: Antibacterial activity of *P. reticulatus*

	Zone of Inhibition in mm				
Organisms	Pet. ether	Chloroform	Acetone	Methanol	Aqueous
E. coli	5	11	15	13	10
P. aeruginosa	-	12	9	8	-
K. pneumoniae	-	9	11	5	9
B. subtilis	5	8	-	-	-

ACKNOWLEDGEMENT

The authors are thankful to St.Xavier's College management for providing infrastructure, constant support and encouragements.

Financial Support: The author T. Renisheya Joy Jeba Malar is thankful to Department of Science and Technology, Govt. of India for providing financial assistance through DST-INSPIRE Fellowship (Ref. No. IF110640).

REFERENCES

Anjenenlu, A.S.R., Jagonmohon R. and Subramanyam C. 1973. Isolation and structural elucidation of three new lignans from the leaves of *Phyllanthus niruri* Linn. *Tetrahedron*. 29:1291.

Bibitha, B., Jisha, V.K., Salitha C.V., Mohan, S., Valsa, A.K., Antibacterial activity of different plant extracts. Short Communication. *Indian J. Microbiol.* 42:361-363, 2002.

De-Lucca, A., Cleveland, T., Rajasekara, K., Boue, S., Brown, R., 2005. Fungal properties of CAY-1, a plant saponin, for emerging fungal pathogens 45th Interscience Conference in Antimicrobial Agents and Chemotherapy. F-490, pp. 180.

Harborne, J.B., 1998. Phytochemical methods: A guide to modern techniques of plant analysis. 3rd edn. Chapman and Hall, New York, pp. 1-150.

Hardi, A.L., Uddin, M.I., 2004, Seasonal variation in the intestinal bacterial flora of hybrid Tilapia (*Oreochromis niloticus* x *Oreochromis aureus*) culture in earthen pond in Saudi Arabia. *Aquaculture.* 229(124):37- 44.

Joshi, K.C., Singh, P., Mehta, A., 1991. Crystalline components of the roots of *Phyllanthus reticulatus*. *J. Ind. Chem. Soc.* 58: 102.

Kambizi, L., Afolayan, A.J., 2008. Extracts from *Aloe ferox* and *Withania somnifera* inhibit *Candida albicans* and *Neisseria gonorrhoea*. *African J. Biotechnol*. 7: 12-15.

Kirtikar, K.R., Basu, B.D., 2003. Indian Medicinal Plants. Dehradun: International Book Distributors, 3060.

Lakshmi Narasimhudu, C., Venkata Raju, R.R., 2012, Phytochemical constituents of Phllanthus species(Euphorbiaceae) From Eastern Ghats of Andhra Pradesh, India, *IRJP*. 3(5):184-200.

Maghrani, M., Zeggwah, N., Michel, J., Eddouks, M., Antihypertensive effect of *Lepidium sativum* in spontaeneously hypertensive rats. *J. Ethnopharam*. 102(1-2):193-197, 2005.

Manach, Mazur, 2005. Bioavailability and bio-efficacy of polyphenols in humans. *Am. J. Clin. Nutr*. 81: 230S-242S

Mandal, P., Sinha Babu, S.P., Mandal, N.C., 2005. Antimicrobial activity of Saponins from *Acacia auriculiformis*. *Fitoterapia*. 76(5):462-565.

Mandal,V., Mohan, Y., Hemalatha, S., 2007,Microwave assisted extraction–an innovative and promising extraction tool for medicinal plant research, *Pharmacog.Rev*. 1: 7-18.

Manjunatha, B.K., 2006. Antibacterial activity of *Pterocarpus santalinus*. *Ind. J. Pharm. Sci.* 68(1):115-116.

Mukherjee, K.L., 2004. Medical Laboratory Technology. Tata McGraw Hill Publishing Company Ltd, New Delhi.

The Wealth of India, 2005. National Institute of Science Communication and Information Resources. Council of Scientific and Industrial Research, New Delhi, pp: 34.

URL, 2012a. http://en.wikipedia.org/wiki/Alkaloids

Vaghisiya,Y., Dave, R., Chanda, S., 2011, Phytochemical analysis of some medicinal plants from western region of India. *Research Journal of Medicinal Plant*. 5(5): 567-576.

Yoshida, T., Seno, K., Takama, Y., Okanda, T., 1982. Tannins and related polyphenols of Euphorbiaceous plants.*Phytochemistry*. 21:1180.

7

Antimicrobial Activity and Phytochemical Screening of *Smilax china* and *Anacyclus pyrethrum*

R. Beena, D. Brindha and V.N. Satakopan

1. INTRODUCTION

Nothing in the universe is non medicinal everything has its value in its own way. Infectious diseases are an important health hazard all over the world, both in developing and developed countries. Several synthetic antibiotics are employed in the treatment of infectious and communicable diseases. A number of researchers nowadays are working seriously to find out substitutes for antibiotics as they cause side effects on the functioning of different parts of the body organs and systems. Antibiotics are also known to disturb the natural intestinal micro flora, thus depriving the benefits of these microbes to human body. Medicinal plants exhibit antimicrobial activity since they contain innumerable biologically active chemical constituents. The medicinal plant wealth, the strong and time tested traditional medicare systems like ayurveda and the human resource base are the strengths of India. Over the last forty years intensive efforts have been made to discover clinically useful antibacterial/ antifungal drugs (Sashikumar *et al.,* 2003). Over 50% of all modern clinical drugs are of natural product origin (Ates, 2003). Higher plants as sources of medicinal compounds have continued to play a dominant role in the maintenance of human health since ancient times (Farombi, 2003).

Plants with possible antimicrobial activity should be tested against an appropriate microbial model to confirm the activity and to ascertain the parameters associated with it. The effect of plant extracts on bacteria has been studied by a very large number of researchers in different parts of the world (Reddy *et al.,* 2001). Medicinal plants are the local heritage with global importance. Medicinal plants are boon of nature to cure a number of ailments to human beings (Bushra Begum and Gangadevi, 2003). Plants have been an essential part of human society since the civilization started (Kunjal Bhatt *et al.*, 2003). Much work has been done on ethno medicinal plants in India (Negi *et al.*, 1993). It has been suggested that aqueous and ethanolic extracts from plants used in allopathic medicine are potential sources of antiviral, antitumoral and antimicrobial agents (Chung *et al.*, 1995). Multidrug resistance among enteropathogens in various geographic regions presents a major threat in the control of infectious diseases. Bacterial infections are commonly seen in number of immunosuppressed and immunocompramised patients. This makes to search for new antibacterial compounds in medicinally important plants. Particular aspect of using medicinal plant as a remedy or home cure for infectious diseases is applied in present study. The aim of the present investigation is to screen the antimicrobial activity of certain medicinally important plant extracts and to screen the phytochemicals present in plant extract both qualitatively and quantitatively with the following objectives; preliminary phytochemical screening of selected medicinal plants, quantification of phytochemicals and investigation of antimicrobial activity against pathogenic bacteria.

2. MATERIALS AND METHODS

2.1. Collection and Storage of Plant Material

The plant material used in the present study namely *Smilax china (paranki Cherkai) and Anacyclus pyrethrum (Akrakara)* were collected from local Ayurvedic shop, Coimbatore. The roots of these plants were used. The roots were cut into small pieces and pulverized by a mechanical grinder. The powdered material was stored in a separate well closed container and labeled.

2.2. Preparation of Plant Extract for Phytochemical Analysis

The plant material may be subjected to preliminary phytochemical screening for the detection of various plant constituents. The plant extract required for the phytochemical analysis was prepared through successive solvent extraction. 50gms of the air-dried powdered plant material was successively extracted with 150ml of 50% ethanolic solvent. The extract dissolved in the solvent was allowed to stand for 24 hrs, filtered through Whatmann.No.1 filter paper and the filtrate to be used for the phytochemical analysis was stored at 4°C.

2.3. Qualitative Analysis of Phytochemicals in the Plant Extract

These studies were performed according to the published standard method (Kokate, 1994).

2.3.1. Detection of alkaloids

A small portion of the solvent free ethanolic extract was stirred with a few drops of diluted hydrochloric acid and filtered. The filtrate was tested with various alkaloidal reagents such as Mayer's reagent (cream precipate), Dragendroff's reagent (orange brown precipate) and Wagner's reagent (reddish brown precipate).

2.3.2. Detection of flavanoids

A small portion of solvent free ethanolic extract was stirred with a few drops of diluted hydrochloric acid and filtered. To the filtrate small pieces of zinc or magnesium were added and the solution was boiled for few minutes. In the presence of flavanooids, reddish pink or dirty brown colour was produced.

2.3.3. Detection of carbohydrates

Fehling's test: To 2ml of ethanolic extract, 1ml of a mixture of equal parts of Fehling's A and Fehling's B was added. The contents were boiled for a few minutes. Formation of red or brick precipate indicated the presence of carbohydrates.

Benedict's test: To 0.5ml of ethanolic extract, 5ml of Benedict's reagent was added and boiled for 5 minutes. Formation of a bluish green colour showed the presence of carbohydrates.

2.3.4. Detection of glycosides

A small amount of ethanolic extract was dissolved in 1ml of water and then aqueous NaOH solution was added. Formation of a yellow colour indicates the presence of glycosides.

2.3.5. Detection of saponins

1ml of ethanolic extract was diluted with 20ml of distilled water and shook well in a graduated cylinder for 15 minutes. A one centimeter layer of foam indicated the presence of saponins.

2.3.6. Detection of phenols:

Ferric chloride test: To 1ml of ethanolic extract few drops of 10% aqueous ferric chloride was added. Formation of blue or green or violet colour indicated the presence of phenols.

Lead acetate test: 1ml of ethanolic extract was diluted to 5ml with distilled water and to it few drops of 1% aqueous solution of lead acetate was added. A yellow precipate indicated the presence of phenols.

2.3.7. Detection of proteins

Million's test: To a small quantity of the ethanolic extract, 5-6 drops of million's reagent was added. A white precipate, which turned red on heating, indicated the presence of proteins.

Biuret test: To 1ml of ethanolic extract, 5-8 drops of 10% NaOH solution followed by one to two drops of 3% copper sulphate was added. Formation of red or violet colour indicated the presence of proteins.

2.3.8. Detection of tannins

Ferric chloride test: To 1-2ml of ethanolic extract few drops of 5% aqueous ferric chloride was added. A bluish black colour, which disappeared on addition of a few ml of diluted.sulphuric acid followed by the formation of yellowish brown precipitate, indicated the presence of tannins.

Lead acetate test: In a test tube containing about 5ml of ethanolic extract, few drops of 1% solution of lead acetate was added. A yellow or red precipitate was formed, indicating the presence of tannins.

2.4. Quantitative analysis of the phytochemicals in the plant extract:

These studies were performed according to the published standard method (Sadhasivam and Manickam, 1994).

2.4.1. Estimation of phenols (folin- ciocalteau method)

Weighed exactly 0.5-1gm of the sample and grinded in a pestle and mortar with 10 times volume of 80% ethanol. Centrifuged the homogenate at 10000 rpm for 20 minutes. The supernatant was saved. The residue was re-extracted with 5 times the volume of 80% ethanol, centrifuged and the supernatant was pooled. The supernatant was evaporated to dryness and the residue was dissolved in a known volume of distilled water (5ml).0.2ml of the extract was pipetted out into a clean dry test tube and the volume was made up to 3ml with water. 0.5ml of folin's ciocalteau reagent was added. After 3 minute, 2ml of sodium carbonate solution was added. Mixed thoroughly and the tube was placed in a boiling water for exactly one minute, cooled and the absorbance was measured at 650nm against a reagent blank.

2.4.2. Estimatin of tannins (folin- denis method)

Tannin- like compounds reduces phosphotungsto-molybdic acids in alkaline solution to produce a highly colored blue solution, the intensity of which is measured in a spectrophotometer at 700nm. Weighed 0.5gm of the powdered material and transferred to a 250ml of conical flask. Added 75ml of water. Heated the flask gently and boiled for 30 minutes. Centrifuged at 2000rpm for 20 minutes and the supernatant was collected in 100ml volumetric flask and the volume was made up. Transferred 1ml of the sample extract to a 100ml volumetric flask containing 75ml water. To this added 5ml of folin-Denis reagent, 10ml of sodium carbonate solution and diluted to 100ml with water. Mixed well and the absorbance was read at 700nm after 30 minutes. A blank was prepared with water instead of sample.

2.5. Preparation of plant extract for antibacterial activity

The plant may be considered a biosynthetic laboratory, not only for the chemical compounds such as carbohydrates, proteins and lipids that are utilized as food by man, but also for a multitude of compounds like glycosides, alkaloids, tannins etc., that exert a physiological effect. A systematic study of a crude drug embraces through consideration of both primary and secondary metabolites derived for the antibacterial activity. 50gm of air-dried powdered plant material was successively extracted with 150ml of 50% ethanolic solvent. The extract is allowed to stand for 24 hours, filtered through Whatmann.No.1 filter paper and the filtrate was evaporated to dryness at room temperature. The dried filtrate was then dissolved in known volume of the particular solvent and stored at 4°C until used for antibacterial assay.

2.6. Test microorganisms

The test microorganisms *viz., Escherichia coli, Klebsiella pneumonia, Proteus* species, *Pseudomonas aeruginosa, Citrobacter, Staphylococcus aureus, Salmonella* species. were obtained from Microbiology Lab, PSG College of Arts and Science. The cultures were maintained on nutrient agar plates at 4°C.

2.7. *In vitro* antibacterial assay (Disc diffusion method)

The disc diffusion method was used to evaluate the antibacterial activity (Hiremath *et.al.,* 1993). Nutrient agar was prepared in the plates as the media for the test microorganisms. Sterile filter paper discs (whatmann.no.1) was impregnated with 50 microlitre of each of the extract{200mg/ml} to give a final concentration of 10mg/ml and left to dry under laminar air flow cabinet for few minutes. The bacterial inoculum was spread evenly onto the surface of the nutrient agar plates using sterile cotton but before the extract discs were positioned on the inoculated agar surface.

Chloramphenicol and *sulphamethizole* (10µg/disc) was used as standard to confirm that all the microorganisms tested were inhibited by the antibiotic. All the plates were incubated for 24 hours at 37°C. The antibacterial activity was interpreted from the size of the diameter of inhibitory zone measured to the nearest millimeter (mm) as observed from the clear zones surrounding the disc.

3. RESULTS AND DISCUSSION

3.1. Qualitative analysis of phytochemicals

The different qualitative preliminary phytochemical tests were performed in selected medicinal plants for establishing profile of given extract for its chemical composition. The tests mentioned in Table 1 were performed on ethanolic extract to identify various phytoconstituents present in them. The results of *S. china* and *A. pyrethrum* were shown in Table 1 and the presence or absence of phytochemicals are indicated by plus (+) or minus (-) signs respectively.

Table 1: Qualitative phytochemical screening of selected ethanolic plant extracts

S.No.	Phytochemical tests	*S. china*	*A. pyrethrum*
1.	**Alkalloids**:		
	i. Dragendroff's Test	++	-
	ii. Wagner's Test	++	-
	iii. Mayer's test	++	-
2.	**Flavanoids**	++	+
3.	**Saponins**	-	+
4.	**Carbohydrates**		
	i. Fehling's Test	+++	++
	ii. Benedict's Test	+++	++
5.	**Proteins**		
	i. Million's Test	-	-
	ii. Biuret Test	-	-
6.	**Phenols**		
	i. Ferric chloride Test	+++	+++
	ii. Lead acetate Test	+++	+++
7.	**Gycosides**	++	++
8.	**Tannins**		
	i. Ferric chloride Test	+++	++
	ii. Lead acetate Test	+++	++

+ : Opaque but no visible particles and no precipate, ++ : Definite turbidity, +++ : Precipitation. - : Absence.

3.2. Quantative analysis of phytochemicals

The phenolic content in *S. china* and *A. pyrethrum* were depicted in Figure 1. From the result it was observed that *S. china* showed maximum content of phenols when compared to *A. pyrethrum*.

The tannin content in *S. china* and *A. pyrethrum* were shown in Figure 2. From the result it was observed that *S. china* showed maximum content of tannins when compared to *A. pyrethrum*. Similar trend was observed by Loganathan (2007). The presence of phenolic and tannin content in *S. china* and *A. pyrethrum* indicate that some bioactive molecules are present in the medicinal plants which might be responsible for curing diseases

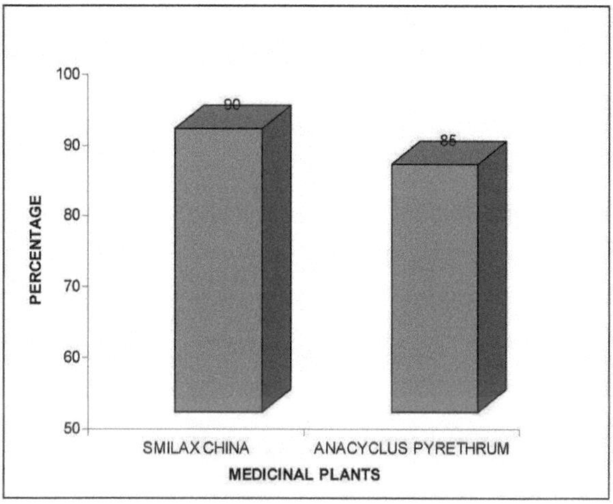

Fig. 1: Levels of phenol content in selected medicinal plants

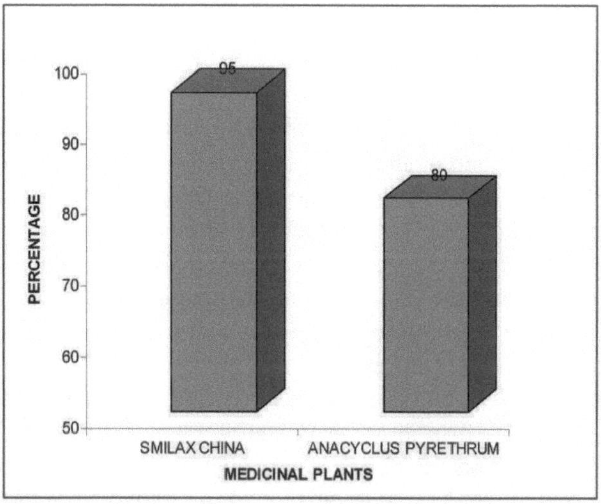

Fig. 2: Levels of tannin content in selected medicinal plants

3.3. Screening for antibacterial assay

The ethanol extract of *S. china* and *A. pyrethrum* were tested for antibacterial activity against various test microorganisms using Disc Diffusion method. The antibacterial activity were expressed as maximum if inhibitor zone is > 20mm, medium if inhibitor zone is >10mm and moderate if inhibitor zone is <10mm (Hiremath, *et.al.*, 1993). The results were depicted in Table 2. *S. china* and *A. pyrethrum* showed maximum zone of inhibition against *coli* and *aeruginosa*. Similar results were observed by Mahalakshmi (2007). *S. china* and *A. pyrethrum* showed medium zone of inhibition against *Klebsiella pneumonia*, *aureus* and *Salmonella* species. Similar results were observed by Deore and Khadabadi (2007).

S. china and *A. pyrethrum* showed moderate zone of inhibition against *Citrobacter* and *Proteus* species. Our results were in good accordance with Deore and Khadabadi, (2007). Amongst the 7 bacterial strains investigated *coli* and *aeruginosa* were the most resistant. It is evident that Gram negative bacteria were more resistant than Gram positive bacteria.

Table 2: Inhibitory gradient by ethanol extract against various tested organisms

Organisms	*Smilax china*	*Anacyclus pyrethrum*
Escherichia coli	+++	+++
Klebsiella pneumoniae	++	++
Pseudomonas aeruginosa	+++	+++
Proteus species	+	+
Citrobacter	+	+
Staphylococcus aureus	++	++
Salmonella species	++	++

Inhibition gradient: + = < 10mm, ++ = > 10mm, +++ = > 20mm.

4. CONCLUSION

The present study reveals clearly that the selected medicinal plants contain maximum, medium and moderate antibacterial activity against both gram negative and gram positive organisms. Besides this, the plant species selected for study also indicated the presence of phytochemical constituents both qualitatively and quantitatively. From the above results it can be concluded that plant extracts have great potential as antimicrobial compounds against microorganisms and that they can be used in the treatment of infectious diseases caused by resistant microorganisms. When compared to *A. pyrethrum, S. china* constitutes rich phytochemical constituents and showed maximum antibacterial activity and hence this plant can be used to discover bioactive natural products that may serve as leads for the development of new pharmaceuticals that address hither to unmet therapeutic needs. Such screening of various natural organic compounds and identifying active agents is the need of the hour, because successful prediction of lead molecule and drug like properties at the onset of drug discovery will pay off later in drug development.

REFERENCES

Ates, D.A. 2003. Antimicrobial activities of some plant extracts used in folk medicine, *Turkish Journal of Biology* 27: 157-162.

Bushra begum, N.R., Gangadevi, T. 2003. Antibacterial activity of seaweeds from south west coast of India, Asian *Journal of Microbial Biotechnology*. 5(3): 319-322.

Chung, T.H., Kim, J.C., Kim, M.K. 1995. Investigation of Korean plant extracts for potential phytotherapeutic agents against *B-virus Hepatitis*. Phytotherapy Research, 9: 429-434.

Deore, Khadabadi. 2007. Invitro antimicrobial studies, *Asian Journal of Microbial Biotechnology*: 807-809.

Farombi, E.O. 2003. African indigenous plants with chemotherapeutic potentials and biotechnological approach to the production of bioactive prophylactic agents, *African Journal of Biotechnology* 2: 662-671,

Hiremath, shrushailapp badami, Swamy, H.K.S., birodar. 1993. Antimicrobial activity of various extracts of *Acalypha Indica, Indian Journal of Microbiology*, 75-77.

Kokate, C.K. 1994. *Practical Pharmacognosy*, pp 245.

Kunjal Bhatt, R., Sonal, J., Gokani, B., Snehal Bagatharia, Vrinda Thaker. 2003. Antimicrobial activities of some medicinal plants: comparison of methods employed and plants studied. *Asian Journal of Microbial Biotechnology*, 5: 455-462.

Loganathan, V. 2007. Preliminary Phytochemical and Antibacterial Screening of *Vitex Negundo Linn., Asian Journal of Microbial Biotechnology, Phytomedica*, 5: 69-72.

Mahalakshmi. 2007. Screening of antibacterial activity, *Asian Journal of Microbial Biotechnology*, 1049-1052.

Negi, K.S., Tiwari, J.K., Guar, R.D. 1993. Notes on ethnobotany of five districts of Garhwal Himalaya, Uttar Pradesh, *India. Ethanobotany* 5: 73-81.

Reddy, P.S., Jamil, K., Madhusudhan, P. 2001. Antibacterial activity of isolates from *Piper longum* and *Taxus baccata. Pharmaceutical Biology*, 58: 1325-1357.

Sadhasivam, S., Manickam, A. 1994. Biochemical methods, New Age International Publishers, pp 270.

Sashikumar, J.M., Remya, M., Janaradhanan, K. 2003. Antibacterial activity of ethnomedicial plants of Nilgris biosphere reserve and western plants, *Asian Journal of Microbial Biotech*, 5(2); 183- 185.

8

Evaluation of Antioxidant and *in vitro* free Radical Scavenging Activity of *Terminalia Bellirica*

Mary Shoba Das, C., Manila, T.N. and Gayathri Devi, S.

1. INTRODUCTION

Reactive oxygen species (ROS) is a collective term used for oxygen containing free radicals, depending on their reactivity and oxidizing ability. They participate in a variety of chemical reactions with biomolecule (Kunwar *et al.,*2011).These free radicals include superoxide anions (O2-), hydroxyl (OH-), peroxyl (ROO), peroxynitrite (ONOO) and nitric oxide (NO) radicals; produced through oxidative process within the body. They can damage both the structure and function of a cell membrane in a chain reaction leading to degenerative disease conditions such as alzheimer, ageing process, cataracts, acute liver toxicity, cardiovascular disease, arteriosclerosis, nephritis, diabetes mellitus, rheumatism, inflammation process and DNA damage that can lead to carcinogenesis (Ghosh *et al.,* 2011).

An antioxidant is a molecule capable of inhibiting or preventing the oxidation of molecules. Antioxidants are substances that may protect cells from the damaging effects of oxygen radicals, highly reactive chemicals that play a part in atherosclerosis, some forms of cancer and reperfusion injuries. Medicinal plants are an important source of antioxidants. The secondary metabolites like phenolics and flavonoids from plants have been reported to be potent free radical scavengers. They are found in all parts of plants such as leaves, fruits, seeds, roots and bark (Chanda and Dave, 2009).

Terminalia bellirica is a native plant of India belonging *Combretaceae* family. It is a large deciduous buttressed tree with thick brownish grey bark having shallow longitudinal fissures. The fruits are thermogenic, anti-inflammatory, styptic, narcotic, digestive, expectorant, ophthalmic, antipyretic, antiemetic and rejuvenating, cough, bronchitis, strangury, skin diseases, leprosy, fever, ulcers and general debility. The mature and dry fruits are useful in diarrhoea and dysentry. Leaf of *Terminalia bellirica* has a promising angiogenic property. The bark is useful in the treatment of anemia and leucoderma. The oil obtained from the seed is trichogenous and is useful in dyspepsia (Warrier, 1996). In this present study the anitoxdiants and *in vitro* free radical scavenging activity of *Terminalia bellirica* was evaluated.

2. MATERIALS AND METHODS

2.1 Collection and Identification of Plant Material

The leaves and fruits of *Terminalia bellirica* was collected from in and around Kannur district, Kerala. The taxonomic identity of the plant was duly authenticated by Botanical Survey of India, Coimbatore, Tamil Nadu. The collected leaves and fruits of the plant were then washed and air dried in the shade at room temperature for complete drying. The plant material was pulverized and used for further investigation.

2.2 Chemicals

Hydrogen peroxide, potassium hydrogen phosphate, potassium hydrogen buffer, ferrous chloride, ferric chloride, ferric cyanide, hydrogen peroxide, ferrous ammonium sulfate, ethylene diamine tetra-acetic acid (EDTA) disodium salt, N- (1-naphthyl) ethylene diaminedihydrochloride, riboflavin, sodium nitroprusside, 1, 1 - diphenyl - 2 - picryl - hydrazyl (DPPH), potassium persulfate were obtained from Himedia and Merck. All other chemicals and solvents used were of analytical grade.

2.3 Quantification Assays

2.3.1 Determination of catalase activity (Luck, 1974)

Catalase is an enzyme, which is present in most cells and catalyzes the decomposition of H_2O_2 to water and oxygen and thereby protects the cells from oxidative damage of H_2O_2 and $OH^{\bullet}$. Homogenized the sample in a blender with M/15 phosphate buffer (pH 7.0) at 4°C and centrifuged. Stirred the sediment with cold M/15 phosphate buffer (pH 7.0), allowed to stand in the cold with occasional shaking and then repeated the extraction once or twice. The combined supernatants were used for the assay. Read against a control cuvette containing the enzyme solution as in the experimental cuvette, but containing H_2O_2 free M/15 phosphate buffer (pH 7.0). Pipetted into the experimental cuvette 3.0 ml of H_2O_2-phosphate buffer (0.16 ml of H_2O_2 made upto 100 ml with M/15 phosphate buffer, pH 7.0). Mixed in 0.01- 0.04 ml sample and noted the time it required for a decrease in absorbance from 0.45 to 0.4. Calculated the concentration of H_2O_2 using the extinction coefficient 0.036 per mmole per ml.

2.3.2 Estimation of peroxidase activity (Reddy et al., 1995)

The activity of peroxidase was determined by the method of Reddy *et al.* (1995). Macerated one gram of the sample with 5 ml (w/v) 0.1M phosphate buffer (pH 6.5) in a homogenizer. Centrifuged the homogenate at 300 g for 15 min. used the supernatant as the enzyme source. All the procedures were carried out at 0-5°C. Pipetted out 3ml of 0.05 M pyrogallol solution and 0.5 to 1.0ml of enzyme extract in a test tube. Adjusted the spectrophotometer to read '0' at 400 nm. Added 0.5ml of 1% hydrogen peroxide in the test cuvette.Recorded the change in the absorbance every 30 seconds upto 3 minutes.

2.3.3 Determination of superoxide dismutase activity (Misra and Fridovich, 1972)

Superoxide anion is known to inactivate enzymes and initiate the damaging chain reactions of lipid peroxidation. Cellular defense mechanisms against superoxides include a series of linked enzyme reactions which remove the toxic radicals and repair the radical induced damage. The first of these enzymes is superoxide dismutase that converts superoxide anion to hydrogen peroxide (Evans and Halliwell, 2001).

The incubation medium contained in a final volume of 3.0 ml, 50 mM potassium phosphate buffer (pH 7.8), 45 mM methionine, 5.3 mM riboflavin, 84 mM NBT and 20 mM KCN. The tubes were placed in an aluminium foil-lined box maintained at 25°C and equipped with 15W fluorescent lamps. Reduced NBT was measured spectrophotometrically at 600 nm after exposure to light for 10 minutes. The maximum reduction was evaluated in the absence of the enzyme. One unit of enzyme activity was defined as the amount of enzyme causing 50 % inhibition of the reduction of NBT.

2.3.4 Estimation of polyphenol oxidase (Esterbauer et al., 1977)

The activity of polyphenol oxidase was estimated by the method of Esterbauer *et al.* (1974). Ground about 5g of the plant sample and made upto 20ml with the medium containing 50mM Tris-HCl (pH 7.2), 0.4M sorbitol and 10mM NaCl. Centrifuge the homogenate at 2000rpm for 10 minutes and used the supernatant for the assay.

Added 2.5ml of 0.2M phosphate buffer (pH 6.5), 0.3ml of catechol solution (0.01 M) into the cuvette and set the spectrophotometer at 495nm. Now added the enzyme extract (0.2ml) and started recording the change in absorbance for every 30 seconds upto 5 minutes

2.3.5 Determination of glutathione peroxidase activity (Reddy et al., 1995)

Peroxidases catalyze the oxidation of a variety of electron donors with the help of H_2O_2 and thus scavenge the endogenous H_2O_2. Pipetted out 3.0 ml of pyrogallol solution (0.05M pyrogallol in 0.05M phosphate buffer, pH 6.5) and 0.05 to 0.1 ml of enzyme extract in cuvettes.Added 0.5 ml of 1 per cent H_2O_2 in the test cuvette.Recorded the change in absorbance every 30 seconds upto 3 minutes at 430 nm.

2.3.6 Estimation of ascorbic acid (Roe and Kuether, 1953)

Ascorbic acid is an electron donor, a reductant and a free radical scavenger. One gram of the sample was homogenized in 10 ml of 4 per cent TCA. Centrifuged at 2000 rpm for 10 minutes. The supernatants obtained were treated with a pinch of activated charcoal. Shaken well and kept for 10 minutes. Centrifuged and 0.5 and 1.0 ml aliquots of this supernatant were taken for the assay. The assay volumes were made upto 2.0 ml with 4 per cent TCA. The working standard solution containing 20-100 mg of ascorbate were pipetted out, the volumes of which were also made upto 2.0 ml with 4 per cent TCA. Added 0.5 ml of 2 per cent DNPH (2,4-dinitrophenyl hydrazine in 9 N H_2SO_4) reagent to all the tubes, followed by 2 drops of 10 per cent thiourea solution. Incubated at 37°C for 3 hrs. The osazones formed were dissolved in 2.5 ml of 85 per cent sulphuric acid, in cold, drop by drop, with no appreciable rise in temperature. To the blank alone, DNPH reagent and thiourea were added after the addition of sulphuric acid. After incubation for 30 minutes at room temperature, the absorbance was read spectrophotometrically at 540 nm. Calculated the content of ascorbic acid in the sample using the standard graph.

2.3.7 Estimation of α-tocopherol (Emmerie-Engel method, as described by Rosenberg, 1992)

Tocopherol, the fat soluble vitamin functions as a scavenger of free radicals to prevent their peroxidative effects on unsaturated lipids of membrane. It was estimated by Emmerie-Engel method, as described by Rosenberg (1992). The samples were weighed (2.5 g) and added 50 ml of 0.1N sulphuric acid slowly without shaking. Stoppered and allowed to stand overnight. The next day, the contents of the flask were shaken vigorously and filtered through Whattman No.1 filter paper, discarding the initial 10-15 ml of the filtrate. Aliquots of the filtrate were used for the estimation. Into three stoppered centrifuge tubes (test, standard and blank) pipetted out 1.5 ml of extract, 1.5 ml of the standard (D, L-a-tocopherol 10 mg/L in absolute alcohol, 91 mg of a-tocopherol is equivalent to 100 mg of tocopherol acetate) and 1.5 ml of water respectively. To the test and blank added 1.5 ml of ethanol and to the standard, added 1.5 ml of water. Added 1.5 ml of xylene to all the tubes, stoppered, mixed well and centrifuged. Transferred 1.0 ml of xylene layer into another stoppered tube. Added 1.0 ml of 2, 2'-dipyridyl reagent

(1.2 g/L n-propanol) to each tube, stoppered and mixed. Pipetted out 1.5 ml of the mixture into spectrophotometer cuvettes and read the extinction of test and standard against the blank at 460 nm. Then, in turn, beginning with the blank, added 0.33 ml of ferric chloride solution. Mixed well and after exactly 15 minutes read the test and standard against the blank at 520 nm.

2.3.8 Estimation of vitamin A (Bayfield and Cole, 1980)

Vitamin A was estimated by the method of Bayfield and Cole (1980). Aliquots of standards ranging from (0-7.5 μg) were pipetted into clean dry test tubes and volume of all the tubes was made upto 0.1 ml with chloroform. To this 2.0 ml of TCA was added rapidly and mixed the contents. Samples were also treated similarly and the absorbance was recorded immediately at 620 nm.

2.3.9 Estimation of polyphenols (Malick and Singh, 1980)

The total phenolic content of the plant sample used in the study was estimated using the method proposed by Malick and Singh (1980). 1g of sample was homogenized using 20ml of 80% ethanol. The homogenate was centrifuged at 10,000rpm for 20 minutes. The supernatant was saved. The residue was reextracted with 10ml of 80% ethanol, centrifuged and collected the supernatant and evaporated to dryness. The residue was dissolved in a known volume of distilled water (50ml) and 2.0ml was taken for the experiment. A working standard of 0.5 – 2.5ml catechol solution corresponding to 50 - 250μg of catechol were pipetted out into a series of test tubes. The volume was made upto 2.5ml with water. To all the tubes added 0.5ml of diluted Folin ciocalteau reagent. After 3 minutes, added 2.0ml of 20% Na_2CO_3 solution to each tube and mixed thoroughly.

The tubes were placed in a boiling water bath for exactly one minute. Cooled and measured at 650nm against a reagent blank. Constructed a standard graph by plotting the concentration of catechol on X-axis and absorbance on Y-axis. From the graph, the amount of polyphenols present in the sample was estimated and expressed as mg of polyphenols per g of the sample.

2.3.10 Estimation of reduced glutathione (Moron et al., 1979)

One gram of the sample was homogenized in 5 per cent TCA to give a 20 per cent homogenate. The precipitated protein was centrifuged at 1000 rpm for 10 minutes. The homogenate was cooled on ice and 0.1 ml of the supernatant was taken for the estimation. The volume of the aliquot was made upto 1.0 ml with 0.2M sodium phosphate buffer (pH 8.0). 2 ml of freshly prepared DTNB solution (0.6 mM in 0.2 M phosphate buffer, pH 8.0) was added to the tubes and the intensity of the yellow colour formed was read at 412 nm in a spectrophotometer after 10 minutes. The standards (2-10 nmoles GSH in 1.0 ml of 5 per cent TCA) were also treated in a similar manner.

2.4 Free radical scavenging activity of the extracts of *Terminalia bellirica*

2.4.1 Extraction of Plant material

The powder obtained from the leaves and fruits of the *Terminalia bellirica* were extracted with suitable solvents using soxhlet apparatus until complete elution. The solvents used in sequential order were acetone, chloroform, ethanol and aqueous. The extracts obtained were concentrated under vacuum on a rotary evaporator at 40°C and all of the dried extracts were stored at "20°C prior to assays of *in vitro* free radical scavenging activity.

2.4.2 DPPH free radical scavenging activity (Mensor et al.*, 2001)*

The antioxidant activity of the extracts, on the basis of the scavenging activity of the stable 1, 1-diphenyl-2-picrylhydrazyl (DPPH) scavenging activity was measured by the method of Mensor *et al.,* 2001.Free radical scavenging activity was determined by 2, 2-diphenyl- 1 -picryl - hydrazyl (DPPH) method by spectrophotometry and compared with the ascorbic acid.A methanolic solution of DPPH (100 mM, 2.95 ml) was added to different concentration of test samples (10-50 mg/ml) and allowed to react at room temperature for 30 minutes. Methanol served as the blank and DPPH in methanol without the extracts served as the positive control. The methanolic solution of ascorbic acid was taken as standard. After 30 minutes, the absorbance was measured at 518 nm. The per cent reduction in the absorbance was calculated from the initial and final absorbance at each level. Concentration of extracts required for 50 per cent reduction was calculated from the calibration curve of the concentration of the extracts Vs per cent reduction in absorbance.

2.4.3 Inhibition of nitric oxide generation (Green and Hill, 1984)

Aqueous solution of sodium nitroprusside spontaneously generates nitric oxide (NO) at physiological pH, which interacts with oxygen to produce nitrite ions. The reaction mixture containing 0.3 ml 100mM sodium nitroprusside, 2.68ml PBS (0.88 per cent NaCl, 0.02 per cent KCl, 0.02 per cent KH_2PO_4 and 0.115 per cent Na_2HPO_4, pH 7.2), and 20 ml of the extracts (20 mg extract) was incubated at 25°C for 15 minutes. Control tubes (100 per cent generation) were prepared without the extracts. After incubation, 0.5 ml of the reaction mixture was removed and 0.5 ml of Griess reagent (1 per cent sulphanilamide, 2 per cent H_3PO_4 and 0.1 per cent naphthalenediaminedihydrochloride) was added to it. The absorbance of the chromophore formed, indicative of the quantum of NO generated, was read at 546 nm. The percentage inhibition of nitric oxide generation was measured by comparing the absorbance values of control and those of test samples.

2.4.4 Determination hydrogen peroxide scavenging radicals (Ruch et al., 1989)

The ability of the plant extract to scavenge H_2O_2 was determined according to the method of Ruch *et al.*, 1989. A solution of H_2O_2 (4mM) was prepared in phosphate buffer (pH 7.4) concentration was determined spectrophotometrically from its absorption at 230 nm was determined after 10 minutes against a blank solution containing phosphate buffer without H_2O_2.

The scavenging activity of H_2O_2 by plant extract and the standard compounds was calculated using the formula.

$$\% \text{ Scavenging activity} = \frac{A_0 - A_1}{A_0} \times 100$$

A_0 – Absorbance of control

A_1- Absorbance of sample at 230 nm

2.4.5 Determination of inhibition of in vitro superoxide generation (Mc Cord and Fridovich, 1969)

The assay tubes contained test sample (20mg concentration) with 0.2ml of ethylene diamine tetra-acetic acid (EDTA), 0.1ml nitro blue tetrazolium (NBT), 0.05ml riboflavin and 2.55ml of phosphate buffer. The control tubes were also set up without leaf extracts. The initial optical densities of the solution were recorded at 560nm. After that, these tubes were placed in an area where they received uniform illumination for 30 minutes.

Again the optical density was measured at 560nm. The difference in optical density before and after illumination is the quantum of superoxide production and the percentage of inhibition by the test sample was calculated by comparing with the optical density of control.

A= O.D at 535 nm

E= Extinction coefficient of MDA

L= Length of cuvette (1cm).

2.4.6 Determination of hydroxyl radicals scavenging activity (Elizabeth and Rao, 1990)

The reaction mixture contained deoxyribose (2.8 mM), $FeCl_3$(0.1 mM), EDTA (0.1 mM), H_2O_2 (1 mM), ascorbate (0.1 mM), KH_2PO_4 – KOH buffer (20 mM, pH 7.4) was taken in a set of clean test tubes and final volume was made upto 1.0 ml. To this 0.5 ml each of 1 per cent trichloroacetic acid, 1 per cent thiobarbituric acid and 500 ml of various concentrations (10 to 50 mg/ml) of the extracts were added. Concentration of extracts required for 50 per cent reduction in absorbance was calculated from the calibration curve of concentration of the extracts Vs percent reduction in absorbance. Copper sulphate solution of 0.1 mM was used as reference standard.

2.5 Statistical Analysis

The results were expressed as mean ± standard deviation (SD) for all the parameters and were statistically analyzed using student's t test. The data was considered as significant or not.

3. RESULTS AND DISCUSSION

Medicinal plants play a key role in the human health care. The traditional medicine refers to a broad range of ancient natural health care practices including folk/tribal practices as well as Ayurveda, Siddha, Amchi and Unani. These medical practices originated from time immemorial and developed gradually, to a large extent, by relying or based on practical experiences without significant references to modern scientific principles. These practices incorporated ancient beliefs (Kumar *et al.,* 2011).

T. bellirica is one of the most popular medicinal plants used in treatment of various human diseases. There is no detailed articles are found about activity of *T. bellirica*. Therefore it has been evaluated for its antioxidants, *in vitro* free radical scavenging activity to get a remedy for a variety of ailments.

Antioxidant activity of *T. bellirica*

Antioxidants provide protection for living organisms from damage caused by uncontrolled production of reactive oxygen species (ROS) and the concomitant lipid peroxidation, protein damage and DNA strand breaking. Current interest is focused on the potential role of antioxidants and antioxidant enzymes in the treatment and prevention of atherosclerosis, heart failure, neurodegenerative disorders, aging, cancer, diabetes mellitus and several others diseases (Majumdar *et al.,* 2010).

3.1 Enzymic antioxidants

All aerobic organisms have evolved with an antioxidant defence system to counteract oxidative stress from ROS (Zou *et al.,* 2008). Endogeneous antioxidant defence has both enzymic and non enzymic components. Enzymic components assessed include catalase, peroxidase, superoxide dismutase, polyphenol oxidase and glutathione peroxidase.

3.1.1 Catalase, Peroxidase and Superoxide dismutase

Catalase, peroxidase and superoxide dismutase activities of *Terminalia bellirica* are presented in Fig. 1.

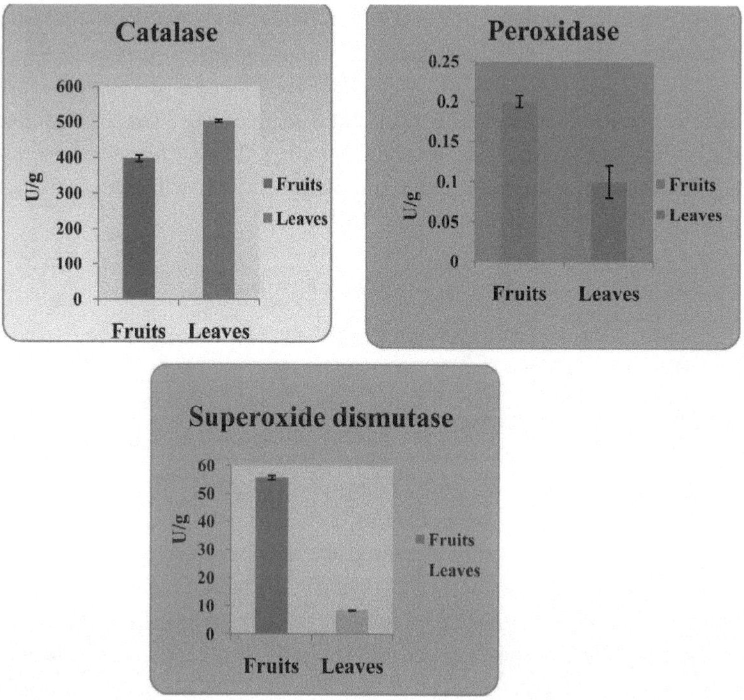

Fig. 1 : Catalase, Peroxisase and Superoxide dismutase activity in the fruits and leaves of *T. bellirica*

Catalase is present in the peroxisome of aerobic cells and is very efficient in promoting the conversion of hydrogen peroxide to water and molecular oxygen (Rahman, 2007). Catalase and hydroperoxidase enzymes convert hydrogen peroxide and hydroperoxides to non radical forms and function as natural antioxidants in human body.Plants posses a protective system composed of antioxidant such as peroxidase and catalase. Catalase is primary H_2O_2scavenger in the peroxisomes and the mitochondria. (Zolfaghari *et al*., 2010). In this study, the fruits and leaves of *T. bellirica* was found to exhibit397.3±9.244, 501.8±42.079 U/g of catalase activity. A marked increase in the catalase activity was observed in the leaves of *T. bellirica*. The present study indicates that *T. bellirica* has considerable catalase activity in the fruits and leaves.

Peroxidase was consideredto be the key enzyme for the decomposition of H_2O_2, especially under catalase inactivitation. Peroxidases, ubiquitously existing in nature, are heme-containing enzymes that oxidize a wide variety of organic and inorganic substrates by reducing H_2O_2 and peroxides. Peroxidases (POD) can form a new cell wall through biosynthesis. In the large family of peroxidases, the process of detoxification of H_2O_2 is particularly important (Aki *et al*., 2000).Our study indicates that fruit (0.2±0.007) of *T. bellirica* had highest activity of peroxidase compared to that of leaves (0.1±0.02 U/g).

From Figure 1, it was cleared that the fruits of *Terminalia bellirica*(55.7±0.707U/g) had maximum superoxide dismutase activity compared to leaves (8.4±0.099U/g) respectively.Superoxide dismutase (SOD), this is one of the most effective intracellular enzymtic antioxidants and it catalyzes the conversion of superoxide anions to dioxygen and hydrogen peroxide. Superoxide dismutase exists in several isoforms, which differ in the nature of active metal center, amino acid composition, co-factors and other features (Rahman, 2007).

3.1.2 Polyphenol oxidase and glutathione peroxidase

Figure 2 presents the activity of polyphenol oxidase and glutathione peroxidase in the fruits and leaves of *T. bellirica*.

The activity of polyphenol oxidase was found to be 0.004±0.014 U/g in the fruits and 0.01±0.002 U/g in the leaves of *T. bellirica.* The leaves show relatively high polyphenol oxidase activity than fruits. Polyphenol oxidase (PPO) is a stress marker enzyme, produced as protective measure against diseases. The antioxidant activity of polyphenols is mainly due to their redox properties, which allow them to act as reducing agents, hydrogen donators, and singlet oxygen quenchers. Antioxidants inhibiting the oxidation of organic molecules are very important, not only for food preservation, but also for the defence of living systems against oxidative stress (Raghuveer *et al.,* 2011).

Glutathione peroxidase is an important antioxidant enzyme that can detoxify hydrogen peroxide and lipid hydroxides using glutathione as a reducing agent. It is clear from the Figure 2 that the glutathione peroxidase activity in leaves of *T. bellirica* (4.5± 0.42426) showed increased activity compared to that of fruits (3 0±.1415).Glutathione peroxidase (GPxs) serve to detoxify products of lipid peroxidation formed due to activity of ROS. GPx decomposes peroxides to water (or alcohol) while simultaneously oxidizing GSH. GPx competes with catalase for H_2O as a substrate and is the major source of protection against low levels of oxidative stress (Ahmad *et al.*, 2008).

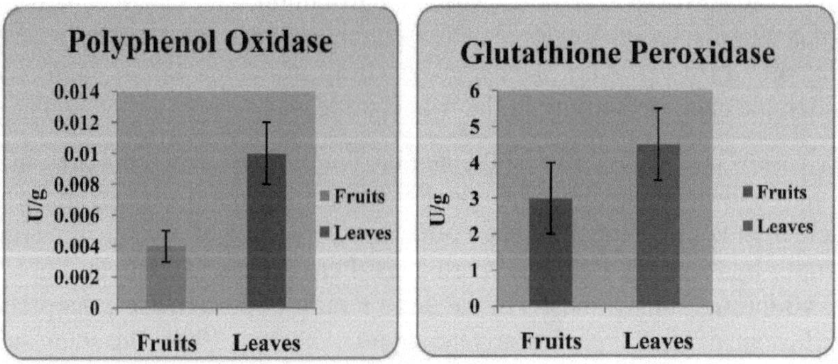

Fig. 2: Polyphenol oxidase and glutathione peroxidase activity in the fruits and leaves of *T. bellirica*

3.2 Non enzymic antioxidants

Apart from enzymic antioxidants, spectrums of non enzymic antioxidants are also important in cellular system in curtailing ROS. The levels of the non enzymic antioxidants were assessed and the results are presented Figure 3.

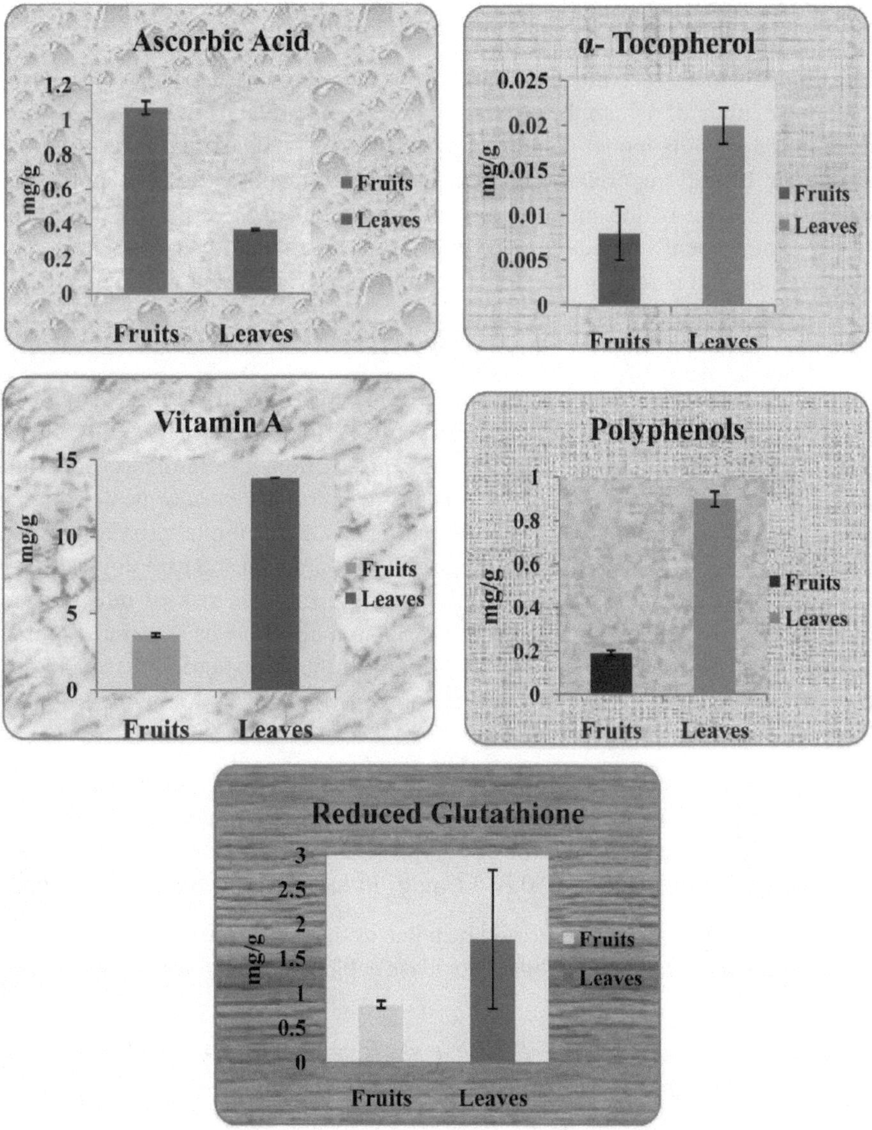

Fig. 3 : Levels of the non enzymic antioxidants in *T. bellirica*

Ascorbic acid, as an antioxi-dant, detoxifies H_2O_2, which is formed by the dismutation of O_2. Several enzymic antioxidants to counteract O2, which is produced by the Mehler reaction and photorespiration. Ascorbic acid is also believed to detoxify $1O_2$ and OH.Ascorbic acid is an antioxidant, which can scavenge O_2 and H_2O_2 non-enzymatically and takes part in scavenging of H_2O_2. Ascorbic acid decreased the activity of these enzymes may be by elimination of free radicals. It also plays a protective role against ROS (Dolatabadian *et al.,* 2009).The present investigation recorded that the ascorbic acid content of *T. bellirica* were found to be1.07±0.035mg/g in fruits and 0.386±0.007mg/g in the leaves.

The figure shows that the fruits and leaves of *T. bellirica* had 0.008±0.009, 0.02±0.022mg/g of α-tocopherol content respectively. The fruits of the *T. bellirica* were found to contain poor content of α-tocopherol. α-Tocopherols, better known as vitamin E is an important antioxidant in the cell membrane have reported that vitamin E is one of the most important antioxidant and least toxic of all lipid soluble antioxidant vitamins. It scavenges peroxyl radical intermediates in lipid peroxidation and is responsible for protecting polyunsaturated fatty acids. It also participates in the detoxification of ROS together with the hydrophilic antioxidants, glutathione and ascorbate (Nirmala *et al.,* 2011).

Vitamin A is a fat soluble micronutrient. It can be derived from animal sources or plant sources (in fruits and vegetables) as provitamin A. It is required for normal growth and development, playing a part in reproduction, differentiation of cellular epithelium, regulation of cell division, genetic regulation and enhancement of immune response. Vitamin A deficiency, a major public health nutritional problem in many developing countries, resulting in a range of developmental and pathological stages with consequence ranging from exophthalmia to increase risk of infection and mortality (Edem, 2009).The present investigation also reveals that the vitamin A content of *T. bellirica* to be 3.59±0.114mg/g in the fruits and 13.83±0.00283in the leaves. The leaves were the richest source of vitamin A compared to that of fruits.

Polyphenols are group of highly hydroxylated phenolic compound present in the extractive fraction of plant samples, which shows a rich antioxidant property to eliminate free radicals. The level of polyphenols in *T. bellirica* was found to be 0.19±0.013 in fruits and 0.90±0.035 (mg/g) in leaves respectively.

The glutathione status of a cell could be taken as the most accurate single indicator of the health of the cell; glutathione peroxidase depletion determines the vulnerability to oxidant attack.

Reduced glutathione level in *T. bellirica* was found to be 0.84±0.053 in fruits and 1.78±1.406 (mg/g) in leaves respectively.

Hence the results suggests that both the leaves and fruits of *Terminalia bellirica* shows good antioxidant content and therefore *Terminalia bellirica* may reduce the risk of serious diseases caused by reaction oxygen species like cancer, aging and diabetes mellitus.

3.3. *In vitro* free radical scavenging activity in the fruit and leaves of *T. bellirica*

Oxidative stress is an important contributor to the pathophysiology of a variety of pathological conditions including cardiovascular dysfunction, atherosclerosis, inflammation, carcinogenesis, drug toxicity, reperfusion injury and neurodegenerative diseases. Plants (fruits, vegetables and medicinal herbs) contain a wide variety of free radical scavenging molecules, such as phenolic compounds, nitrogenous compounds, vitamins, terpenoids and some other endogenous metabolites, that are rich in antioxidant activity. Scavenging of free radicals by antioxidants could reduce the fibrosis process in the tissue. They are considered as possible protective agents reducing oxidative damage to the human body and are naturally abundant in fruits and are able to neutralize free radicals donating an electron and converting them to harmless molecules (Kaneria *et al.,* 2009).

3.3.1 DPPH radical scavenging activity of T. bellirica

1,1-Diphenyl-2-picryl hydrazyl (DPPH) is a stable nitrogen centered free radical which can be effectively scavenged by antioxidants (Villano *et al.,* 2007). DPPH scavenging activity of the extracts of *T. bellirica* is presented in Figure 4.

From the above figure, it is clear that the extract of *T. bellirica* have got markedly increased action against DPPH radicals. The ethanol extract in fruit showed

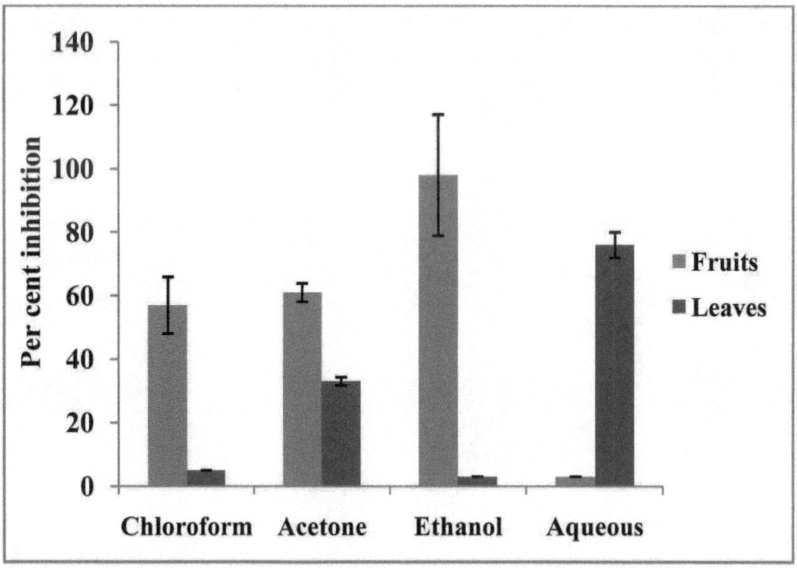

Fig. 4: DPPH radical scavenging activity of the fruits and leaves of *T. bellirica*

maximum inhibitory activity (98.6 per cent). Aqueous extract in leaves (76.01 per cent) showed relatively high inhibition against DPPH radical tested. The order of DPPH radical scavenging ability of different extracts of fruits and leaves was ethanol > aqueous > acetone > chloroform, acetone > chloroform > ethanol> aqueous respectively.

In this study, the ethanol extract of fruits and aqueous extract of *T. bellirica* were found to have strong DPPH radical scavenging activity. Hence, the present data suggests that fruits and leaves extract of *T. bellirica* are capable of scavenging free radicals significantly. Thus they may be able to prevent the initiation and propagation of free radical mediated chain reaction by stabilizing reactive species via electron or hydrogen donation before such deleterious reaction can occur.

3.3.2 Inhibition of nitric oxide generation

Nitric oxide (NO) and its derivative peroxynitrate ($ONOO^-$), produced from nitric oxide and superoxide were the main source of RNS. It is an important inters and intracellular regulatory molecules or multiple physiological functions and is enzymatically synthesized via the oxidation of the terminal guanidine nitrogen atom of L-arginine by a family of nitric oxide synthase (NOS), which are either constitutive (cnos) or inducible (Inos). The damage from NO during the inflammation or carcinogenic process might be decreased by the NO scavengers and NOS enzyme inhibitors. A number of phytochemicals, such as quercetin, tocopherol and catechin have been found to inhibit the damage of nitric oxide. The per cent inhibition of nitric oxide generation is shown in Figure 5.

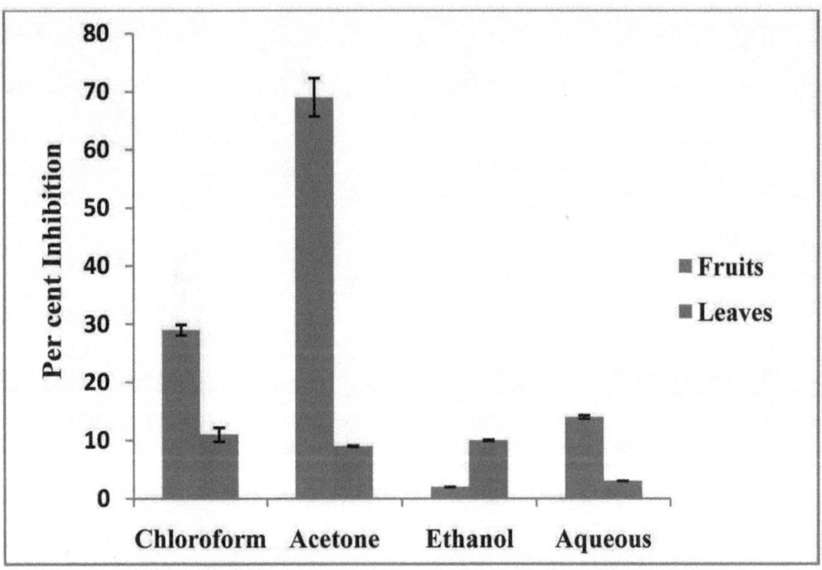

Fig. 5: Inhibition of nitric oxide generation of *T. bellirica*

It is clear from the figure that the per cent inhibition of different extracts of fruits and leaves of *T. bellirica* showed relatively moderate free radical scavenging activity, which was shown in the following order: acetone > chloroform > aqueous > ethanol, chloroform > ethanol > acetone > aqueous.

3.3.3 H_2O_2 scavenging activity of T. bellirica

Unsaturated lipids in cell membrane are susceptible to peroxidation. Consequently, these chain reactions initiated by hydroxyl radical attacking lipids are extended by the generated lipid hydroperoxide free radicals. Despite the low reactivity of hydrogen peroxide, the high penetrability of cellular membrane might lead to hydroxyl reaction formation when it reacts with ferrous ion or superoxide anion radical in the cell leading to oxidative stress (Guoi *et al.*, 2006).

Therefore, in this study, in order to analyze the antioxidant property of different extracts of fruits and leaves of *T. bellirica,* it is necessary to determine the H_2O_2 scavenging of *T. bellirica*.

The results of the hydrogen peroxide scavenging ability of fruits and leaves of *T. bellirica* in Figure 6

The fruits and leaves of *T. bellirica* showed moderate percentage of inhibition against hydrogen peroxide, where compared to extracts of fruits and leaves of *T. bellirica* showed significant increase in H_2O_2 scavenging activity. The order of inhibition of fruits and leaves were acetone > ethanol > chloroform > aqueous, chloroform > acetone > aqueous > ethanol.

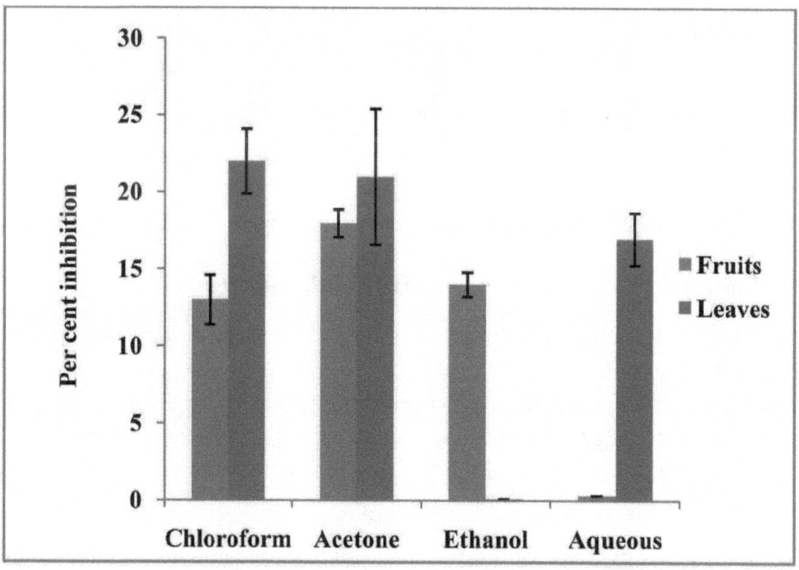

Fig. 6 : H_2O_2 scavenging activity of *T. bellirica*

3.3.4 Inhibition of superoxide scavenging activity of *T. bellirica*

Superoxide anion radical is one of the strongest reactiveoxygen species among the free radicals that are generated (Olayinka and Okoh, 2010). Superoxide can act either as oxidant or reductant. It can oxidize sulphur, ascorbic acid or NADPH and it can reduce cytochrome C and metal ions (Kumar, 2011). A number of phytochemicals, such as tocopherol and catechins have been found to inhibit the damage of superoxide. The per cent inhibition of superoxide generation is shown in Figure 7

This present study reveals that the aqueous extract of leaves shows significantly high scavenging activity than that of fruits. The order of inhibition of fruit and leaves are as follows: acetone > aqueous > chloroform > ethanol, aqueous > chloroform > acetone > ethanol.

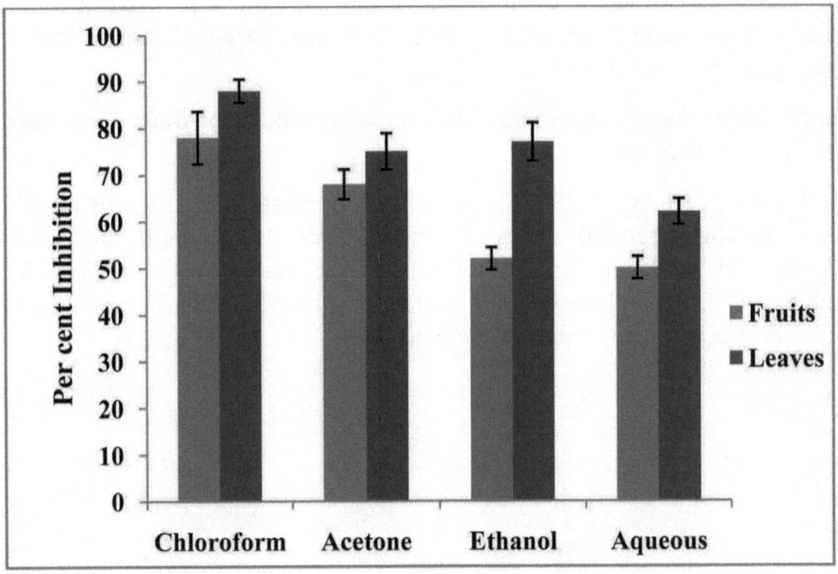

Fig. 7: Super oxide scavenging activity of *T. bellirica*

3.3.5 Hydroxyl radical scavenging activity of T. bellirica

The hydroxyl radical reacts with fettour ion or superoxide anion radical in the cell leading to oxidative stress.The hydroxyl radical is an extremely reactive free radical formed in biological systems and has been implicated as a highly damaging species in free radical pathology, capable of damaging the biomlecules of living cells (Zhang *et al.,* 2009). Figure.8: Presents the hydroxyl radical scavenging activity against *T. bellirica*

The above figure reveals that the per cent inhibition of different extract of fruits and leaves of *T. bellirica* in the following order: chloroform > aqueous > ethanol

> acetone, aqueous > acetone > chloroform > ethanol respectively. The aqueous extract of leaves of *T. bellirica* exhibits shows strongest inhibitory activity when compared to other extracts.

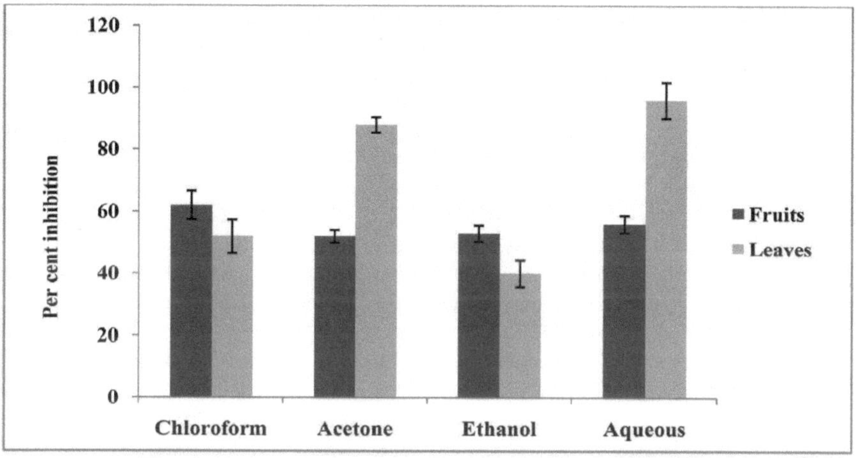

Fig. 8: Hydroxyl radical scavenging activity of *T. bellirica*

T. bellirica was found to exhibit significant *in vitro* free radical scavenging activity. It may reduce the risk of diseases caused by free radicals. Hence, this study show the capability of the fruits and leaves extracts to scavenge free radicals indicating that they may be useful therapeutic agents for radical related pathological damage. Thus *T. bellirica* likely to be a potential medicinal plant resource for developing antioxidants.

4. CONCLUSION

From the present study it is evident that the fruit and leaf extracts of *T. bellirica* possess good antioxidant and hence able to scavenge *in vitro* free radicals effectively. The plant can be further explored to be used as a potent therapeutic agent for free radical related pathological damage.

REFERENCES

Ahmad, P., Sarwat, M., Satyawati, S., 2008. Reactive Oxygen Species, Antioxidants and Signaling in Plants.*Journal of Plant Biology* 51 (3): 167-173.

Aki, C.,Güneysu, E.,Acar, O., 2009.Effect of industrial wastewater on total protein and the peroxidase activity in plants.*Afr. J. Biotechnol.*, 8 (20): 5445-5448.

Bayfield, R.F., Cole, E.R., 1980.Colorimetric estimation of vitamin A with trichloroacetic acid.*Meth. Enzymol.* 67: 189-203.

Chanda, S., Dave, R., 2009. *In vitro* models for antioxidant activity evaluation and some medicinal plants possessing antioxidant properties: an overview. *African Journal of Microbiology Research.* 3: 981-996.

Dolatabadian, A.,Jouneghani, R.S., 2009. Impact of Exogenous Ascorbic Acid on Antioxidant Activity and Some Physiological Traits of Common Bean Subjected to Salinity Stress.Not. Bot. Hort. *Agrobot. Cluj.* 37 (2): 165-172.

Edem, D.O., 2009. Vitamin A: A Review.Asi. J. *Clini. Nutr.* 1(1): 65-82.

Elizabeth, K.,Rao, M.N.A., 1990.Oxygen radical scavenging activity of curcumin. *Int. J. Pharm.*, 58: 237-240.

Esterbauer, H., Schwartz, Z.,Hayan, M., 1977.A rapid assay for catechol oxidase and laccase using 2-nitro-5 thio benzoic acid. *Anal. Biochem.*, 489: 489-494.

Evans, P.,Halliwell, B., 2001.Micronutrients: oxidant/antioxidant status. *Brit. J. Nutr.*, 85: S67-S74.

Ghosh, T., Maity, T.K., Sengupta, P., 2011. Antidiabetic and in vivo Antioxidant Activity of Ethanolic Extract of BacopamonnieriLinn. Aerial Parts: A Possible Mechanism of Action. *Iranian Journal of Pharmaceutical Research* 7 (1): 61-68.

Green, M.J., Hill, H.A.O., 1984. Chemistry of dioxygen *Methods.Enzymol.*, 105.

Guoi, Z., Yuan, Y., Ye, T., Xiang, J., Zhao, G., 2006.Antioxidant activities of *Salvia miltiorrhiza*and *Panaxnotoginseng*. *Food Chem.* 99: 767-774.

Kaneria, M., Baravalia, Y., Vaghasiya, Y., Chanda, S., 2009. Determination of Antibacterial and Antioxidant Potential of Some Medicinal Plants from Saurashtra Region, India. *Indian. J. Pharm. Sci.* 71(4): 406–412.

Kumar, H.C., Ramesh, A., Kumar, S.N.J., Ishaq, M.B., 2011. A Review on Hepatoprotective Activity of Medicinal Plants.*IJPSR*.. 2(3): 501-51.

Kumar, S., 2011. Free radicals and antioxidants, Human and food system. *Ad. App. Sci. Res.* 2(1): 129-135.

Kunwar, A., Priyadarsini, K.I., 2011.Free radicals, oxidative stress and importance of antioxidants in human health.*J. Med. Allied. Sci.* 1(2): 53-60.

Luck, H., 1974.Methods of enzymatic analysis.Academic Press, 885-894.

Majumdar, S., Bhattacharya, S., Haldar, K.P., 2010. Comparative *in vitro* Free Radical Scavenging Activity of some Indigenous Plants. *International Journal of Pharm Tech Research.* 2(2): 1046-1049.

Malick C.P., Singh M.B., 1980. In plant enzymology and Histo enzymology.Kalyani publishers New Delhi, 286.

Mc Cord, J.M.,Fridovich, I., 1969. The reduction of cytochrome C by milk xanthine oxidase. J. *Biol. Chem.* 243: 5753-5760.

Mensor, C.C., Menezes, F.S., Leitan, G.G., Reis, A.S., Dossaltos, T.C., Carbe, C.S.,Leitao, S.G., 2001. Screening of Brazilian plant extracts for antioxidant activity by use of DPPH free radical methods.Phytother. *Res.* 15: 127-130.

Misra, M.P., Fridovich, I., 1972.The role of superoxide anion in the autooxidation of epinephrine and simple assay for superoxide dismutase. *J. Biol. Chem.* 247: 31-70.

Moron, M.S., Bepierre, J.W.,Mannerwick, B., 1979.Levels of glutathione reductase and glutathione-S-transferase in rat lung and liver.*Biochem. Biophys.Acta.* 582.

Nirmala, A., Saroja, S., Devi, G.S., 2011.Antidiabetic Activity of *Basellarubra*and its Relationship with the Antioxidant Property.*British Biotechnology Journal.* 1 (1): 1-9.

Olayinka, A. A.,Okoh, A.I., 2010. Preliminary phytochemical screening and *in vitro* antioxidant activities of the aqueous extract of *Helichrysumlongifolium*DC , BMC Comp. Alt. Med. 10: 21.

Raghuveer.,Dindor, P.P., Joshi, A., 2011. A Comparative Physico-Chemical Analysis of ShodhitaAndMaritaHingula. *Ayurvedic Renaissance.* 3-4.

Rahman, K., 2007. Studies on free radicals, antioxidants and co-factors.*Clin. Int. Aging.* 2(2): 219-236.

Reddy, K.P., Subhani, S.M., Khan, P.A., Kumar, K.B., 1995. Effect of light and benzyl adenine on dark-treated growing rice leaves, II changes in peroxidase activity. *Plant Cell Physiology.* 24: 987-994.

Roe, J.H., Kuether, A., 1953. The determination of ascorbic acid in whole blood and urine through 2, 4- dinitro phenyl hydrazine derivative of dehydroascorbic acid. J. *Biol. Chem*. 147: 399-404.

Rosenberg, H.R., 1992. Chemistry and Physiology of the vitamins.Inter Science Publishers, Inc. New York, 452-453.

Ruch, R.J., Cheny, S.J., Klaunig, J.E., 1989. Prevention of cytotoxicity and inhibition of intracellular communication by antioxidant catechinsisolated from Chinese green tea. *Carcinogenesis*. 10: 1003-1008.

Villano, D., Pachona, F.M.S.,Moyab, M.L., Trancosa, A.M., Parrila, G.M.C., 2007. Radical scavenging ability of polyphenolic compounds towards DPPH free radical. *Talanta*. 71: 230-235.

Warriersundaram, 1996.Indian medicinal plants.Shanda Publication, Kerala. 327-330.

Zhang, J., Luo, J., Wan, L.,Nie, G., Zhang, P., Zhang, Z., Zhao, B., 2009. α-Amyloid peptide increases levels of iron content and oxidative stress in human cell and *Caenorhabditiselegans* models of Alzheimer disease. *Free Radical Biology and Medicine*. 50: 122-129.

Zolfaghari, R., S. M. Hosseini, S.M., Korori, S. A. A., 2010.Relationship between peroxidase and catalase with metabolism and environmentalfactors in Beech (*Fagusorientalis*Lipsky), in three different elevations. *Int. J. Environmental Sciences*. 1(2): 243-252.

Zou, Y., Qian, Z., Li, Y., Kim, M., Lee, S., Kim, S., 2008. Antioxidant effects of phlorotannins isolated from *Ishigeikamurae* in free radical mediated oxidative systems. *J. Agric. Food Chem*. 56:7001-7009.

□□□

9

Phytochemical and Antibacterial Activity of the Green Seaweed *Caulerpa scalpelliformis* (R. Br.)Weber-Van Bosse

Babu, A., Patric Raja, D. and Johnson, M.

1. INTRODUCTION

Seaweeds are the extraordinary sustainable resources in the marine ecosystem which have been used as a source of food, feed and medicine. The number of natural products isolated from marine organisms increases rapidly, and now exceeds with hundreds of new compounds being discovered every year (Chapman and Chapman. 1980, Cheennubhotla *et al.,* 1986; Chennubhotla *et al.,* 1987; Faulkner, 2002; Proksch and Muller, 2006). Several products of algal origin such as alginate, carrragenean and agar as phycocolliods have been used for decades in medicine and pharmacy (Fitton, 2006). Seaweeds are considered as source of bioactive compounds and produce a great variety of secondary metabolites characterized by a broad spectrum of biological activities. Seaweeds contain various inorganic and organic substances which can benefit human health (Kuda *et al.,* 2002). The *in vitro* studies on green, brown and red algae confirmed the following potential activities viz., cytostatic, antiviral, antihelminthic, antifungal and antibacterial (Lindequist and Schweder, 2001; Newman *et al.,* 2003; Dhargalkar and Neelam, 2005; Huang *et al.,* 2005; Gonzalez del Val *et al.,* 2001; Espeche *et al.,* 1984).

Caulerpa, highly differentiated coenocytic green algae, has a unique and rapid wound healing mechanism which prevents excessive cytoplasmic loss. In about

15-30 sec after wounding the rhizome, blades or rhizoids, a mass of gelatinous material rapidly accumulates the cut end and develops in to a compact, fibrillar matrix. Another compound from *Caulerpa racemosa*, caulerpin, a cyclooctatetraene ring containing pigment (Ayyad and Badria, 1994; Clavijo *et al.,* 1996) displaying antitumor activity was studied *in vitro* (Ayyad and Badria, 1994). Numerous green algae (Chlorophyta) are especially deterrent to herbivores by producing repulsive and toxic secondary metabolites (Jung *et al.,* 2002). Caulerpenyne is the main secondary metabolite synthesized by the *Caulerpa* green marine algae family (Ulvophyceae, Caulerpales), which played a major role in its chemical defense (Pohnert and Jung, 2003). It inhibits the growth of microorganisms, (Parent-Massin *et al.,* 1996; Pesando *et al.,* 1998; Barbier *et al.,* 2001). With this knowledge, the present study was aimed to study the phytochemical and antibacterial properties of *C. scalpelliformis.*

2. MATERIALS AND METHODS

2.1 Collection of Materials

Caulerpa scalpelliformis (R. Br.) Weber-Van Bosse was collected by handpicking at Rasthacaud, Kayakuamri, Tamil Nadu, India (Lat N 08^{0}08'308" E77^{0}32'80"). The collected samples were cleaned well with seawater to remove all the extraneous matter such as epiphytes, sand particles, pebbles and shells and brought to the laboratory in plastic bags. The collected samples were then thoroughly washed with tap water followed by distilled water. For drying, washed seaweeds were blotted on the blotting paper and spread out at room temperature in shade under dark condition. The shade dried samples were grounded to fine powder using tissue blender. The powdered samples were then stored in refrigerator at 4°C for further use.

2.2. Preparation of extract

2.2.1 Hot Extraction

The dried and powered materials (30g) were extracted successively with 180 ml of petroleum ether, methanol, chloroform, acetone, benzene and water by using Soxhlet extractor for 8 h at a temperature not exceeding the boiling point of the solvent. The extracts were filtered using Whattman filter paper (No.1) and then concentrated in vacuum at 40° C using Rotary evaporator. The residues obtained were stored in a freezer -20° C until further tests.

2.2.3. Preliminary Phytochemical Analysis

Phytochemical screening of *C. scalpelliformis* extracts were carried out by the Harborne method to confirm the presence of steroids, phenolic compounds, alkaloids aminoacids tannins, flavonoids and anthroquinnones.

2.2.4. Antibacterial Activity

The antibacterial potential of *C. scalpelliformis* was examined in using disc diffusion method (Kirby Bauer) against the clinical isolates viz., *Staphylococcus aureus, Salmonella typhi, Bacillus cereus, Escherichia coli, Klebsiella pneumoniae, Proteus mirabilis, Pseudomonous aeruginosa, Enterococcus faecalis* and *Serratia marcesens*. The pathogens were isolated from the clinical samples (Vivek's laboratory, Nagercoil, Tamil Nadu, India) and idenitfied by the conventional method.

3. RESULTS AND DISCUSSION

Preliminary phytochemical screening of eight different chemical compounds (steroids, phenolic compounds, alkaloids, aminoacids, tannins, flavonoids and anthraquinones) were tested in seven different extracts. Thus out of 49 ($7 \times 7 = 49$) tests for the presence or absence of the above compounds, only 22 gave positive results and the remaining gave negative results. The 22 positive results show the presence of steroids, phenolic groups, saponins, tannin, flavonoids. Aminoacids did not show any positive result for their presence in any of the six extracts tested. Tannins shows the maximum presence in six different extracts followed by steroids in 2 extracts. Among the six different extracts, methanolic extract showed the presence of maximum number (6) of compounds. Next to that acetone extracts dispalyed the presence of 2 compounds viz., tannins and terpenoids (Table 1).

The *in vitro* antibacterial activity of *C. scalpelliformis* crude extracts was assessed by the disc diffusion method. The results were compared with the standard antibiotic streptomycin. The methanolic and acetone extracts showed the highest zone of inhibition of >15 mm against *E. coli, K. pneumoniae* and *P. aeruginosa* and moderate activity was observed in methanolic and acetone extract against *B. cereus*. No activity was noticed against *P. mirabilis, S. marcesens* and *S. aureus* of all the tested extracts. Chloroform extract does not show inhibition against any of the tested pathogens.

Table 1: Preliminary Phytochemical studies on *C. scalpelliformis*

Tests	Solvents						
	Pet. ether	Benzene	Chloroform	Acetone	Ethanol	Methanol	Aqueous
Steroids	+	+	+	+	+	+	-
Alkaloids	-	-	-	+	+	+	+
Phenolics	-	-	-	-	+	+	-
Flavonoids	-	-	-	-	-	+	-
Tannin	-	+	+	+	+	+	+
Anthraquinone	+	+	+	-	-	-	-
Aminoacids	-	-	-	-	-	-	-

Table 2: Antibacterial activity of *C. scalpelliformis*

Organism	Methanol	Acetone	Chloroform	Benzene	Petroleum ether	Aqueous	Control
E.coli	20	18	-	10	17	12	21
K. pneumoniae	19	28	-	-	7	10	25
P. aeruginosa	24	15	-	-	-	-	31
P. mirabilis	9	-	-	-	-	-	19
S. typhi	12	-	-	-	-	-	23
S. marcesens	8	-	-	-	-	-	26
S. aureus	-	-	-	-	-	-	16
B. cereus	12	11	-	-	-	-	25
E. faecalis	-	-	-	-	-	-	17

The results of the bioefficacy studies showed higher activity against gram positive and gram negative bacteria than the standard antibiotic disc. The highest actiivty of *C. scalpelliformis* may be due to the presence of steroids, phenolic compounds, alkaloids, aminoacids, tannins, flavonoids and anthraquinones. Phenolic compounds are commonly found in plants, including seaweeds, and have been reported to have a wide range of biological activities including antibacterial and antioxidant properties (Duan *et al.,* 2006; Kuda *et al.,* 2007; Wang *et al.,* 2009). Shan *et al.,* (1992) indicated that the antibacterial activity is due to different chemical agents present in the extract, including flavonoids, triterpenoids and other compounds of phenolic nature. In the present study also methanolic and acetone extracts showed the good antibacterial activity against *E. coli, K. pneumonia* and *P. aeruginosa.* Dar *et al.,* 2007: indicated that the dried methanolic extracts of three brown species of seaweeds were found to have good antimicrobial activity. Govindasamy *et al.,* (2011) reaveled that methanolic extracts of green seaweeds showed more activity than red and brown seaweeds. Ravikumar *et al.,* (2002) confirmed that among the seaweed extracts the acetone extracts of *Caulerpa cupressoides* gives maximum inhibitory activity against the bacterium *E. coli.* The results of present study found that extracts obtained using ethanol were more active than those obtained with hexane. Thus the results of the present study showed that the extracts of *C. scalpelliformis* contained broad spectrum of antibacterial activities due to the presence of various bioactive compounds. Further work is needed to find out the active compounds responsible for biological activity.

REFERENCES

Ayyad, S.E.N., Badria, F.A., 1994. Caulerpin, an antitumor indole alkaloid from *Caulerpa racemosa*, Alex. *J. Pharm. Sci*. 8, 217.

Barbier, P., Guise,S., Huitorel,P., Amade,P.,Pesando,D., Briand,C., Peyrot, V., 2001.Caulerpenynefrom Caulerpa taxifolia has an antiproliferative activity on tumorcelllineSK-N-SHandmodifies the microtubulenetwork.*LifeSci.*70,415–429.

Chapman, V.J., Chapman, D.J., 1980. Seaweeds and their uses, 3rd edn Chapman and Hall, New York. 34- 37.

Cheennubhotla, V.S.K., Kalimuthu, S., Selvaraj, M., 1986. Seaweed culture its feasibility and industries utilization. *Proc. Symp.Coastal Aquaculture*. 4, 1206-1209.

Chennubhotla, V.S.K., Ramachandhu, B.S., Kaladharan, P., Dharmaraj, S.K., 1987. Seaweed resources of Kerala coast seminar on fisheries research and development in Kerala Trivandrum (Abstract).

Clavijo, R.E., Weiss-Lopez, B., Rivera, P., 1996. Vibrational spectra of Caulerpin, a green algae pigment. *Bol Soc Chil Quim* 41, 153.

Dar, A., Baig H.S., Saifullah S.M., Ahmad, V.U., Yasmeen, S., Nizamuddin, M., 2007. Effect of seasonal variation on the anti-inflammatory activity of *Sargassum wightii* growing on the N. Arabian Sea coast of Pakistan. *J. Exp. Mar. Biol. Ecol.* 351, 1–9.

Dhargalkar, V.K., Neelam, P., 2005. "Seaweed: promising plant of the millennium," *Science and Culture*, Vol. 71, pp. 60-66.

Duan, X. J., Zhang, W. W, Li, X. M., Wang, B. G., 2006. Evaluation of antioxidant property of extract and fractions obtained from a red alga, *Polysiphonia urceolata. Food Chemistry* 95: 37-43.

Espeche, M.E., Fraile, E.R., Mayer, A.M.S., 1984. Screening of Argentine marine algae from antimicrobial activity. *Hydrobiologia*. 116/117, 525-528.

Faulkner, D.J., 2002. *Marine natural products*. Nat. Prod. Rep. 19, 1-48.

Fitton, J. H., 2006. Antiviral properties of marine algae. In: Critchley, A. T., Ohno, M. and Largo, D. B. (eds). World Seaweed Resources, Windows & Macintosch. ETI Information Services, Workingham, U.K., p. 7.

González del Val, A., Platas, G., Basilio, A., Cabello, A., Gorrochategui, J., Suay, I., Vicente, F., Portillo, E., Jiménez del Rio, M., Garcia RG, Peláez, F., 2001. Screening of antimicrobial activities in red, green and brown macroalgae from Gran Canaria (Canary Islands, Spain). *Int. Microbial* 4, 35-40.

Govindasamy, C., Narayani, S., Arulpriya, M., Ruban, P., Anantharaj, K., Srinivasan, R., 2011. *In vitro* antimicrobial activities of seaweed extracts against human pathogens, *Journal of Pharmacy Research*,4(7),2076-2077.

Harbourne, J. B. 1994. The Flavonoids: Advances in Research Since 1986. London, UK:

Huang, X., Zhoui, H., and Zhang, H., 2005. "The effect of *Sargassum fusiforme* polysaccharide extracts on vibriosis resistance and immune activity of the shrimp Fenneropenaeus chinensis," *Fish and Shellfish Imunology*, Vol. 20, pp. 750-757.

Jung, V., Thibaut,T., Meinesz,A., Pohnert,G., 2002.Comparison of the wound-activated transformation of caulerpenyne by invasive and noninvasive Caulerpa species of the Mediterranean.*J.Chem.Ecol.*28, 2091–2105.

Kuda, T., Kunii, T., Goto, H., Suzuki, T. and Yano, T., 2007.Varieties of antioxidant and antibacterial properties of *Ecklonia stolonifera* and *Ecklonia kurome* products harvested and processed in the Noto peninsula, Japan. *Food Chemistry* 103: 900-905.

Kuda, T., Taniguchi, E., Nishizawa, M., Araki, Y., 2002. Fate of water-soluble polysaccharides in dried *Chorda filum* a brown alga during water washing. *Journal of Food Composition and Analysis* 15: 3-9.

Lindequist, U., Schweder, T., 2001. Marine Biotechnology. In: Rehm, H.J., Reed, G. (Eds.), Biotechnology, vol. 10. Wiley-VCH, Weinheim, to its growth stages. Indian pp: 441-484. J. Marine multiresistant pathogens. *Annals of Microbiology* 58 (3): 535-541.

Newman, D.J., Cragg, G.M., Snader, K.M., 2003. Natural products as source of new drugs over the period 1981-2002. *J. Natural Products,* 66: 1022-1037.

Parent-Massin, D., Fournier, V., Amade, P., Lemee, R., Durand-Clement, M., Delescluse, C., Pesando, D., 1996. Evaluation of the toxicological risk to humans of caulerpenyne using human hematopoietic progeni- tors, melanocytes, and keratinocytes in culture. J. Toxicol. Environ. *Health* 47, 47–59.

Pesando, D., Huitorel, P., Dolcini, V., Amade, P., Girard, J.P., 1998. Caulerpenyne interferes with microtubule-dependent events during the first mitotic cycle of sea urchin eggs. Eur. *J. Cell Biol.* 77:19–26.

Pohnert, G., Jung, V., 2003. Intracellular compartmentation in the biosynthesis of caulerpenyne: study on intact macroalgae using stable- isotope-labeled precursors. Org. Lett. 5:5091-5093.

Proksch, P., Muller, W.E.G., 2006. Frontiers in Marine Biotechnology. Horizon Bioscience: Norfolk, U.K.species. *Phytochemistry*, 58:291–297

Ravikumar, S., Anburajan.L., Ramanathan, G., Kaliaperumal, H., 2002. Screeening of seaweed extracts against antibiotics resistant post operative infectious pathogens,*Seaweed Res Utilin* ., 24(1): 95-99.

Renuka Bai, N., 2010. Evaluation of *Gracilaria Fergusonii* for Phytochemical analysis and antibacterial activity. *Plant Archieves*. 10 (2):711-713.

Shan, B., Cai, Y. Z., Brooks, J. D., Corke, H., 2007. The *in vitro* antibacterial activity of dietary spice and medicinal herb extracts. *International Journal of Food Microbiology,* 117: 112-119.

Wang, T., Jónsdóttir, R., Ólafsdóttir, G., 2009. Total phenolic compounds, radical scavenging and metal chelation of extracts from Icelandic seaweeds. *Food Chemistry* 116: 240-248.

10

Paliyar Folklore Health Remedial Practices in Sirumalai Hills, Tamil Nadu, India

Ramesh Kannan N., Harikrishnan K., Vijayalatha K.R. and Natarajan E.

1. INTRODUCTION

Our Indian traditional medicine depends on herbals which always played an important role in the health problems of many countries. In India, the tribal people are using variety of herbals for effective curing of various ailments. These plants are often used by the local people of that area for treatment of various diseases. Ethnomedicinal studies have offered immense value and opportunities for the development of new drugs. Living close to nature, traditional societies have acquired unique knowledge about the use of wild flora and fauna, most of which are unknown to the people who live away from such natural ecosystem as forests and hills. The hills have been the sources of medicinal plants since the time man realized the preventive and curative properties of plants and started using them for human health care.

Tribal people throughout the world, including India, have developed their own cultures, customs, religious rites, taboos, legends and myths, folk tales, medicine, food, etc., They are the repository of accumulated experience and knowledge of the indigenous vegetation; this can be utilized in tribal development. These days greater emphasis is being laid on this traditional knowledge and on using it in the "bioprospecting" of natural products as a new source of food and medicine.

India is one of the twelve megadiversity countries in the world and has 17,000 flowering plants. Among the 25 hotspots in the world, the Eastern Himalayas and the Western Ghats are the two hotspots of India (Kaveriappa and Shetty, 2001).

The country possess a total of 427 tribal communities. The largest proportion of tribes is found in Mizoram (95%), followed by Lakshadweep (93%), Nagaland (88%), Meghalaya (86%) and Arunachal Pradesh (64%) (Dinesh Kumar, 2005; Kala, 2005). Tamil Nadu is situated on the Eastern side of the Indian Peninsula. The state has 29 districts and includes 37 tribal communities. They are distributed over various districts. Three major tribal groups of Tamil Nadu are the Valaya or Valaiyans, who live in the hill tracks of Madurai and Tanjore districts, and the Paliyan and Pulayan, living in the Western Ghats. A considerable number of studies have been published on the ethnobotany and ethnomedicine of many tribal groups in different regions of India. Ethnobotanical studies of two groups of Valaiyans, residing in the Vellimalli hills and the Seithur hills (Rajendran *et al.*, 2001; Ganesan and Kesavan, 2003 respectively) have already been undertaken.

The study was performed in Sirumalai Hills of Dindigul District of Tamilnadu, India, which is one of the oldest continuous inhabited hill of the world and has only a little information on the hill's flora.

Paliyars are good herbalists. They use plants in different forms like juice extracts, decoctions, pastes, infusions, etc. A juice extract is prepared by grinding the cleaned plants or plant parts with water; the extract is used after filtering. A decoction is obtained by boiling the plants or plant parts in water. A paste is made by crushing small parts of a plant with water and making this into a soft mass. An infusion is prepared by soaking the cleaned plant or plant parts in water for a few hours or days; afterwards it is filtered and used.

1.1 Study Area

The area of investigation lies approximately between 10°00' and 10°30' N latitude and 77°33' and 78°15' E longitude of Dindigul district in Tamilnadu. Sirumalai is a small hill station 20 Km away from Dindigul on the Natham Road. The Sirumalai is mentioned in the ancient Ayurvedic books as a place abundant with many medicinal plants. The Siddha form of healing has been developed on these hills by the sages a thousand years back. It is 1,200 metre high and spreads over 60,000 acres.

2. METHODOLOGY

Ethnomedicinal information was gathered by contacting the medicine man, the headman and persons with a thorough knowledge of plants. The field surveys were carried out in Sirumalai Hills of Tamilnadu during November 2008 to March 2009. The information gathered was confirmed by different groups of Paliyar dwelling in different places of the area of investigation. The methodology of previous workers was adopted (Jain and Goel, 1995). The data was meticulously entered in a field notebook. The voucher specimens were collected and identified by referring to standard floras (Gamble and Fischer, 1959; Matthew, 1981; Nair and Henry, 1983; Henry *et al.*, 1987; Henry *et al.*, 1989; Matthew, 1991). All the voucher plants were preserved in the form of herbarium specimens, deposited in the Centre for Research and PG Department of Botany, National College, Trichy, Tamil Nadu - 620001.

3. RESULTS AND DISCUSSION

The tribal and rural population of India is highly dependent on medicinal plants for meeting their healthcare needs. This has attracted the attention of several botanists and plant scientists who directed vigorous research towards the investigation of several medicinal plants; this has resulted in an extensive scientific literature.

The present investigation shows that 78 plant species belonging to 70 genera and 44 families are employed for medicinal purposes by a particular tribal population.

The following 10 plants are employed in fever related problems: *Adiantum philippense, Clerodendrum serratum, Coffea arabica, Coleus aromaticus, Ipomoea hederacea, Cymbopogon citratus, Stachytarpheta indica, Vernonia cinerea* and *Ziziphus jujuba.*

The following 13 plants are used for treating colds and cough: *Coffea arabica, Dodonaea viscosa, Euphorbia pulcherrima, Melia dubia, Opuntia dillenii, Ricinus communis, Rubia cordifolia, Stachytarpheta indica, Syzygium cumini, Vernonia cinerea* and *Vitex negundo. Papver somniferum* mainly used for irritating cough. Among these *Syzygium cumini* especially given for appetite increase.

The following 5 plants are employed for the gynecological problems: *Artocarpus heterophyllus, Asclepias curassavica, Ocimum gratissimum* and *Toona ciliata.*

Asclepias curassavica, Cymbopogon citratus and *Stachytarpheta indica* are employed in the treatment of various problems related to the heart.

Plumbago zeylanica and *Urginea indica* are used for treatment of tumors.

Asparagus racemosus, Crataeva religiosa, Hybanthes enneaspermus and *Tectona grandis* are taken for urinary troubles.

Argemone maxicana, Artocarpus heterophyllus, Achyranthus aspera and *Ocimum americanum* are utilized in the treatment of tooth ache. *Coffea arabica, Alpinia galanga* and *Measa indica* are taken for digestive problems.

Coffea arabica, Ficus glomerata, Mimosa pudica, Stachytarpheta indica, Papaver somniferum, Sesbania aegyptiaca, Vernonia cinerea and *Ziziphus jujuba* are employed in the treatment of diarrhoea.

The following plants are administered for various problems such as asthma (*Adhatoda vasica, Datura metel* and *Adiantum philippense*), leprosy (*Adiantum philippense, Commelina benghalensis* and *Jasminum angustifolium*) and dysentery

A list of medicinal plants with their binomial, family, vernacular name, useful parts and medicinal uses is provided below.

Binomial Name	Family name	Local name	Uses
Abutilon indicum L. Sweet.	Malvaceae	Thuthi.	Used to breathing problems.
Achyranthus aspera L.	Amaranthaceae	Nayuruvi.	Root relieves toothache and leaf used for piles.
Adhatoda vasica Nees	Acanthaceae	Adathodi	Leaf extract is taken to relieve cough and cure asthma.
Adiantum philippense L.	Adiantaceae	Kalijhant.	Decoction of fresh leaves relives cough, asthma, fever, leprosy and hair fall.
Alpinia galanga Sw.	Zingiberaceae	Perarathai.	Slightly burnt rhizome is powdered and mixed with honey that mixture is applied to stomach upset and indigestion in children.
Andrographis echioides Nees.	Acanthaceae	Gopuramthangi.	Leaf powdered and consumed along with rice water used to snake bite and eczema.
Andrographis paniculata Nees.	Acanthaceae	Chiriyanangai	Paste form of entire plant is applied externally for skin disease.
Anisomeles malabarica. R.	Lamiaceae	Peythumbai	Leaf of this plant with the leaf of *Alangium salvifollium* is made into paste and applied to cure chronic wounds.
Argemone maxicana L.	Papavaraceae	Brahmathandu	The seeds are poured in 'mahua oil' and are applied to eczema and itching. Roots are burnt to provide heat in treatment of piles. The smoke of the burning seeds is used to relieve tooth ache.
Aristolochia bracteolata L.	Aristolochiaceae	Aduthinnapalai	Leaf paste applied on the head while bath relieves dandruff and other infections.
Aristolochia indica L.	Aristolochiaceae	Eswara mooligai	Leaf past is used for eczema. Leaf decoction relieves stomach pain during menses.
Artocarpus heterophyllus Lamk.	Moraceae	Jack fruit.	Latex of Panasa plant is applied to glandular swellings and abscesses to promote suppuration. Roots are used for toothache, skin disease, sores and sterility in women.
Asclepias curassavica Linn.	Asclepiadaceae	Raktapushpa	Whole plant is considered to be emetic, styptic and purgative. It has cardiotonic effect. Root is emetic, cathartic and used as an astringent and a remedy in piles and gonorrhoea.
Asclepias gigantea Linn	Asclepiadaceae	Arka	Dried and powdered plant is taken with milk acts as a good tonic.
Asparagus racemosus Wild.	Liliaceae	.	Dried leaves are powdered and are taken orally to cure stomach ache, and urinary disorders

(Contd.)

Binomial Name	Family name	Local name	Uses
Astracantha longifolia Nees	Acanthaceae	Neermulli	The decoction of leaves and curry is made out of leaves are taken in the case of anemia.
Calophyllum inophyllum L.	Clusiaceae	Punnai	Seed oil applied externally in rheumatism and skin affections. A decoction of it employed for indolent ulcers.
Calotropis gigantea R.BR.	Asclepiadaceae	Erukku	Milky latex is applied on the wounds on the legs of livestock.
Cassia fistula L.	Caesalpiniaceae	Sarakkonnai	The decoction of the bark is mixed with garlic and powdered pepper and later on given to cattle as purgative.
Centella asiatica L. urban	Apiaceae	Vallarai	Leaves are used in vein problems, whole plant acts as brain tonic.
Cissus quadrangularis L. Mant.	Vitaceae	Pirandai	Leaves used to cure dysentery and paste of the leaves and chilly mixed with salt and administered for appetite in livestock.
Clerodendrum serratum L.	Verbanaceae	Bhaarangee	Used to cure fever.
Cocculus hirsutus L.	Menispermaceae	Kattukkodi	Leaf Juice is applied to eczema. Roots are used for rheumatism and stomachache in children.
Coffea arabica Linn.	Rubiaceae	Kappi	Diuretic or promotes the flow of urine. Stimulates the flow of digestive juices and also taken for migraine, fever, gout and diarrhoea.
Coleus aromaticus Benth.	Lamiaceae	Indian borage	Karpuravalli with its distinctive smelling leaves is a common home remedy for infantile cough, cold and fever.
Commelina benghalensis L.	Commelianaceae	Kanavazhi	Mucilaginous and starchy rhizomes are cooked and eaten Juice of leaves used in emollient and leprosy.
Crataeva religiosa Forst.	Capparidaceae	Maralingam	Bark stimulated liver, its extract used as laxative promoting appetite and other urinal affections. Flowers are astringent.
Cymbopogon citratus DC Stap F.	Gramineae Poaceae		Whole plant used in heart diseases, throat problems, pain, cough and fever
Datura metel L.	Solanaceae	Karuoomathai	Petals are shade dried and broken into pieces and an instant beedi is prepared and smoked for asthma. Leaf juice and coconut oil mixture in the ratio of 2:1 respectively applied for wound healing.
Dodonaea viscose L. Jacq.	Sapindaceae	Virali	Used to Cold, wounds, intestinal worms.
Eclipta prostrata L. Mant.	Asteraceae	Karichalankanni	Leaves applied to relieve dandruff and blackening gray hair.

(Contd.)

Binomial Name	Family name	Local name	Uses
Elettaria cardamomum Maton	Zingiberaceae	Cardamom	Leaves have carminative and stimulant properties. Commonly used either as adjuvants or correctives of cordial tonic and purgative medicines.
Erythrina indica Lamk	Fabaceae	Indian coral tree	Leaves are laxative, diuretic, anthelmintic, galactgogue, emmenagogue, applied externally for dispersing veneral buboes and for relieving pain in joints.
Euphorbia pulcherrima Willd. ex Klotzsch.		Euphorbiaceae	Palperuchi For stomach pain and dysentery. The leaf paste mixed with salt taken internally relieves stomach pain. Leaf decoction relieves dysentery.
Ficus glomerata Roxb.	Moraceae	Athi	Leaves used in bilious affections. Root used in diarrhea and diabetes. Bark is given to cattle in render-pest disease and decoction is used as a vulnerary.
Gliricidia sapium Jacq	Papilionaceae	Madriado	The extract obtained from its leaves is used to remove external parasites.
Glycyrrhiza glabra L.	Fabaceae	Adhimadhuram	Roots taken orally to relieve throat pain and leaf juice applied for lip cracks.
Gymnema sylvestre R.Br	Asclepiadaceae	Sirukurinchan	Extract of leaves taken orally along with milk for diabetes.
Hemidesmus indicus L.	Asclepiadaceae	Nannari	The root powder is mixed with sugar water and taken internally as a cooling beverage.
Hibiscus rosa – sinensis L.	Malvaceae	Semparuthi	Paste of fresh leaves is applied on head for healthy and black hair.
Hybanthes enneaspermus L. F. Muell.	Violaceae	Orithalthamarai	Root is used in urinary disorders. Leaves used for bowel complaints.
Ipomoea hederacea Jacq.	Convolvalaceae	Kakkat	The seeds are used as laxative, carminative and cure inflammations, abdominal diseases, fever, headache, head problems and bronchitis.
Ipomoea sepiaria Roxb.	Convolvulaceae	Talikkodi	Plant juice is used as deobstruent and diuretic.
Jasminum angustifolium Vahl.	Oleaceae	Kattumullai	Roots are used externally for ringworm and leprosy.
Jatropa curcus L.	Euphobiaceae	Kattamanakku	It is used to paralysis and externally for skin troubles and rheumatism. Leaf juice taken orally along with goat milk for scabies and ringworm.
Mangigera indica L.	Anacardiaceae	Maamaram	The latex from leaf, stem and bark is used to treat head cracks.
Measa indica (Roxb) DC	Myrsinaceae		Used for Indigestion
Melia dubia	Meliaceae	Malaivembu	The fruit is bitter and nauseous, good for abdominal colic, useful as a cholagogue in malarial fevers
Mimosa pudica Linn	Mimosaceae	Garrambi Gardal Gila	Lajjalu is astringent, cooling, antiseptic, alterative and blood purifier. It is used in burning sensation of body, diarrhoea, dysentery, hemophilic conditions. Leucorrhoea, morbid, conditions of vagina.

(Contd.)

Binomial Name	Family name	Local name	Uses
Morinda tinctoria Roxb.	Rubiaceae	Manjanatti	Decoction of leaf is administered for stomach pain and dysentery in children.
Musa paradisiaca Linn.	Musaceae	Banana	Root is anthelmintic. Ash of root or whole plant is anthelmintic. Flowers are astringent. Juice of the stem is used in otalgia and haemolysis.
Ocimum americanum Linn.	Lamiaceae	Kaalli tullasa	A decoction is taken for cough, that of leaves for dysentery, also used as a mouth was for relieving tooth ache. Juice of leaves is given to children for cold catarrh and bronchitis.
Ocimum gratissimum Linn	Labiatae Lamiaceae		Decoction of leaves in useful in seminal weakness and is remedy for gonorrhoea. Seeds are given for headache and neuralgia.
Opuntia dillenii Haw.	Cactaceae "	Sappathikalli.	Fruits edible used in whooping cough. Pulp also applied in opthalmia and control spasmodic cough and expectoration.
Oxalis corniculata L.	Oxalidaceae	Puliyarai	Fresh juice of plant given in piles and anemia. Leaf juice is given to counteract poisoning.
Papaver somniferum L.	Papaveraceae	Kasakasa	Poppy seed oil used in culinary purposes, free from narcotic action and also used in diarrhoea and dysentery. Extract used for irritating cough.
Passiflora edulis	Passifloraceae	Passion fruit	The pulp of the fruit is stimulant and tonic.
Phoenix sylvestris L. Roxb.	Arecaceae	Khajoor	Fruit is tonic and restorative.
Phyllanthus emblica L.	Euphorbiaceae	Nelli.	Fruits used in cooling, diuretic and laxative.
Plumbago zeylanica L.	Plumbaginaceae	Chitur	Root paste with tuber of *Arisaema murrayi* and bone of *Egrella gazelta* (cattle erget) is applied to cure tumor.
Randia spinosa Lam	Rubiaceae	Mindhal	Pulp of fruit, dried and powdered is a valuable emetic. Useful in bronchitis, pain in muscles, paralysis, also used as fish poison.
Ricinus communis Linn.	Euphorbiaceae	Amanakku	The leaf juice is taken orally or washed leaves are tied on the breast to increase the secretion of milk in women. The oil prepared from the seeds is applied on lower stomach to get relief from stomach.
Rubia cordifolia L.	Rubiaceae		Used for Stomach ache.
Sesbania aegytiaca Pers.	Fabaceae	Chithagathi	Leaf decoction used in diarrhoea, itches and skin eruptions.
Sesbania grandiflora Pers.	Fabaceae	Agaththi	Leaf decoction to eliminate worms and cures in the stomach.

(Contd.)

Binomial Name	Family name	Local name	Uses
Solanum erianthum D. Don	Solanaceae	Malai sundai	The ripened or unripened fruits are boiled with water and the vapour is inhaled once or twice a week through mouth.
Solanum surattrense Burm. f.	Solanaceae	Kandankathiri	Fresh or dried fruit are kept in fire and the smoke is inhaled.
Solanum torvum Swartz.	Solanaceae	Sundai	Unripened fruit is cooked and taken internally to eradicate intestinal worms.
Stachytarpheta indica C.B. Clark.	Verbenaceae	Simainaivirunji	Used for ulcers, ophthalmic problems, diarrhoea, dysentery, fevers, rheumatic inflammations, cardiac problems, oedema and stomach ailments.
Streblus asper Lour.	Moraceae	Piraayanmaram	Leaves are used in swelling and ulcers. Latex is used as antiseptic. Seed paste is used externally in leucoderma.
Syzygium cumini L. Skeels	Myrtaceae	Jamun	Used to diabetes, dysentry, headache and increases appetites.
Tectona grandis Linn. f.	Verbenaceae	Teak	Roots are given in anuria and tension of urine. Flowers are useful in bronchitis and urinary discharge.
Toona ciliata Roemer	Melialeae	Samdarnavembu	Bark of is astringent, tonic; anti-periodic, used in chronic infantile dysentery and as external application for ulcers. Flowers are emmenagogue; used in menstrual disorders.
Urginea indica (Roxb.) Kunth	Liliaceae	Kolikanda	Taken for burnings and tumors.
Vernonia cinerea Less.	Asteraceae	Sahadevishanglamir	Herbs used against malaria. Roots used as an antihelimintic, their decoction given in diarrhoea and stomach ache. Leaf juice used in fever, cough and rheumatism.
Vitex negundo L.	Verbenaceae	Nochi	Leavs boiled and the vapour is inhaled to relieve cough, cold body pain and headache.
Withania somnifera Dunal.	Solanaceae	Asvakantha	Root paste is applied externally for inflammatory conditions ulcers and scabies.
Ziziphus jujuba Lam	Rhamanaceae	Bora	Badara root is bitter and cooling, cures kapha, biliousness and headache. Deduction of roots is used in fever and as powder applied to old wounds and ulcers. Bark, a remedy in diarrhoea.

(*Cissus quadrangularis, Euphorbia pulcherrima, Mimosa pudica, Morinda tinctoria, Ocimum americanum, Papaver somniferum, Stachytarpheta indica, Syzygium cumini* and *Toona ciliata*).

The following plants are give as diuretics: *Coffea arabica, Erythrina indica, Ficus glomerata, Gymnema sylvestre, Ipomoea sepiaria. Syzygium cumini* and *Phyllanthus emblica* are given against diabetes. *Adiantum philippense, Eclipta prostrata* and *Hibiscus rosa – sinensis* are administered to prevent hair fall and maintaining black hair. *Anisomeles malabarica, Datura metel, Dodonaea viscosa* and *Ziziphus jujuba are* employed for wounds.

Calotropis gigantea's milk latex is applied on the wounds on the legs of livestock. *Dodonaea viscosa, Sesbania grandiflora* and *Solanum torvum* are used against intestinal worms.

Ipomoea hederacea, Ocimum gratissimum, Syzygium cumini, Vitex negundo and *Ziziphus jujuba* are taken for headache. *Mimosa pudica, Urginea indica* and *Ziziphus jujuba* employed for heart burning sensation of the body.

The following plants are administered for ulcers *Calophyllum inophyllum, Stachytarpheta indica, Toona ciliata, Streblus asper, Withania somnifera* and *Ziziphus jujuba.*

Ipomoea hederacea, Ocimum americanum, Randia spinosa and *Tectona grandis* are given for bronchitis.

The following plants are give for pain: *Crataeva religiosa, Cymbopogon citratus, Randia spinosa, Erythrina indica, Ipomoea hederacea* and *Phyllanthus emblica* are used as laxative herbals.

Ipomoea hederacea and *Elettaria cardamomum* are taken for prevention of gas formation.

Passiflora edulis, Elettaria cardamomum, Phoenix sylvestris are used as stimulant.

Astracantha longifolia leaves are taken in the case of anemia. *Asparagus racemosus, Euphorbia pulcherrima, Morinda tinctoria* and *Vernonia cinerea* are used for stomach pain. *Oxalis corniculata* used for anemia and piles.

Two plants are used to cure breathing problems (*Solanum surattrense* and *Solanum erianthum*). *Argemone maxicana* used for eczema and itching; *Aristololchia indica* is used for eczema and its leaf decoction relieves stomach pain during menses. *Cocculus hirsutus* is applied to eczema. Roots are used for rheumatism and stomachache in children.

Plant are used in the treatment of gout (*Coffea arabica*), Rheumatic (*Stachytarpheta indica, Calophyllum inophyllum, Jatropa curcus* and *Vernonia cinerea*), neuralgia (*Ocimum gratissimum*), malarial fever (*Melia dubia* and *Vernonia cinerea*), piles (*Argemone maxicana, Achyranthus aspera* and *Oxalis corniculata*), glandular swelling (*Artocarpus heterophyllus*).

The following plants are used as a blood purifier (*Mimosa pudica*), milk secretion (*Ricinus communis*), vein tonic (*Centella asiatica*), brain tonic (*Centella asiatica*), hemophilic (*Mimosa pudica*), antiseptic (*Mimosa pudica* and *Streblus asper*) and Cordial tonic (*Elettaria cardamomum*).

Aristolochia bracteolata paste applied on the head while bath, relieves dandruff and other infections.

Cassia fistula- its decoction of the bark is mixed with garlic and powdered pepper and later on given to cattle as purgative.

Glycyrrhiza glabra - roots relieve throat pain and leaf juice applied for cracks on the lips.

Hemidesmus indicus - root powder is mixed with sugar water and taken internally as a cooling beverage. *Phyllanthus emblica* also used for cooling.

Employing one plant only cures the following ailments: External parasites (*Gliricidia sopium*), Ringworm (*Jasminum angustifolium and Jatropa curcus*), Paralysis (*Randia spinosa*), Head cracks (*Mangigera indica),* vomiting (*Melia dubia*).

Musa paradisica used as an antihelmintic agent and *Randia spinosa* used against fish poison; *Andrographis echioides Nees* used for snake bite.

The studies on the ethnomedicinal lore of the Seithur hills (Rajendran *et al.*, 2001) revealed the use of 36 plant species belonging to 33 genera distributed over 24 families of flowering plants by the Valaya tribals of Virudhunagar. Out of these, 7 species were used for scabies and other skin diseases, 2 species for jaundice, 4 species for headache, 3 species for fever and one species for diabetes.

The present study documented 78 medicinal plants from Sirumalai belonging to 70 genera distributed over 44 families. Out of these, 10 species were employed in fever, 11 species in cases of cold and cough, 5 species in cases of gynecological, 3 species in heart problems, 2 species in cases of cancer and urinary problems. Nearly 9 plants are recorded in the present study were also employed by the Valaiya of Seithur hills. The studies conducted on the ethnomedicinal plants used by the Valaiyans of Vellimalai hills (Ganesan and Kesavan, 2003) indexed 84 angiospermic plant species belonging to 28 genera distributed over 40 families for the treatment of various disorders, such as wounds, cuts, stomach pain, diabetes, fever, eczema, dandruff, cold, body heat, poisonous bites. The present study on the ethnobotany of the Sirumalai revealed 78 medicinal plants for curing the above-mentioned diseases.

The ethnobotanical uses of 161 species of Angiospermic plants distributed over 139 genera representing 60 families were reported from the Thottianiackans of Semmalai hills, Tiruchirapalli district (Venkatesan *et al.*, 2005); the uses of 119 plants for medicinal purposes were recorded. In the present study 47 plants out of a total of 85 were categorized as of medicinal value.

The Valaiyans utilize a large number of plants for medicinal and other purposes. A fair number of these species, listed for medicinal purposes by the Valaiyans, were found to be common and already reported from other places (Karuppusamy, *et al.*, 2000; Ganesan and Kesavan, 2003; Ganesan *et al.*, 2004 and Venkatesan *et al.*, 2005).

The uses of certain species have gained importance due to their manifold properties. For example, *Abutilon indicum* and *Achyranthus aspera* are used for piles and leg pain, *Acalypha indica* for eczema and chest pain, *Mollugo nudicaulis* for cold and skin diseases, *Andrographis paniculata, Calophyllum inophyllum, Jatropa curcus* and *Sesbania aegytiaca* for skin disease, *Streblus asper* - seed paste is used externally in leucoderma. *Notonia grandis* for earache and eczema. Sharma and Singh (2001) and Jain and Patole (2001) have also reported similar activities in certain plant species (*Cassia fistula* and *Gloriosa superba*) in their studies on ethnobotanical uses of plants. A few species described here as well as by various other authors from different regions as used by various tribes are common, but used for different purposes.

Information on some very useful medicines known to the tribal or ethnic communities through the experiences of ages is usually passed on from generation to generation. The tribal ways of life are rapidly breaking off under the impact of improved communication facilities and still more under the various tribal welfare schemes of the government (Matthew, 1981). As the tribal population is gradually adopting modern ways life, their heritage of traditional knowledge of plants will soon be lost forever.

For the ethnobotanist it is an urgent task to record for posterity whatever is valuable in the traditions of the tribes, their way of life, and their knowledge of plants before all this has disappeared before the close of this century. At the present this valuable diversity is under serious threat due to habitat destruction, overexploitation, shifting cultivation and several anthropogenic and natural pressures. This suggests that the awareness about medicinal plants and their conservation strategy for the sustainability of our ecosystems has to be stimulated.

REFERENCES

Dinesh Kumar, 2005. Tribal literacy disparity in India. *Curr. Sci.*, 88(5): 676.

Gamble, J.S. and Fischer, C.E.C. 1959. The flora of the Presidency of Madras. Rep. Ed. I–III. Botanical Survey of India, Calcutta.

Ganesan, S. and Kesavan, L., 2003. Ethnomedicinal plants used by the ethnic group Valaiyans of Vellimalai hills (Reserved Forest), Tamil Nadu, India. *J. Econ. Taxon. Bot.* 27(3): 754-760.

Ganesan. S. Suresh. N. and Kesavan. L., 2004. Ethnomedicinal Survey of Lower Palni Hills of Tamil Nadu. *Indian Journal of Traditional Knowledge.* 3(3):299–304.

Henry, A.N. Chithra, V. and Balakrishnan, N.P. 1989. Flora of Tamil Nadu, Series–I, Vol. III Botanical survey of India, southern – Circle, Coimbatore.

Henry, A.N. Kumari, G.R., Chitra, V., 1987. Flora of Tamilnadu, India, Series 1: *Analysis Botanical Survey of India*, Southern Circle, Coimbatore.

Jain, A.K. and Patole, S.N. 2001. Less-known medicinal uses of plants among some tribal and rural communities of Pachmarchi forest (M.P.). *Ethnobotany.* 13: 96–100.

Jain, S.K. and Goel, A.K. 1995. A manual of Ethnobotany, S.K. Jain (Ed.), Scientific Publishers, Jodhpur. 142–153.

Kala, C.P., 2005. Ethnomedicinal botany of the Apatani in the Eastern Himalaya region of India. *Journal of Ethnobiology and Ethnomedicine*, 1(11) (http://www.ethnobiomed.com/contant/1/1/11).

Karuppusamy, S. Rajasekaran, K.M. and Kumuthakalavalli, R., 2000. Needs for Diversity - Conservation of Traditional medicinal Plant Resources of Dindigul District, Tamil Nadu. *Ecol. Env. & Cons.* 6(3): 273–278.

Kaveriappa and Shetty, B.V., 2001. Biodiversity of the Western Ghats with special reference to conservation of plant Diversity at Kaiga. *An International Journal of Nuclear Power*, 15:1 - 4.

Matthew, K.M., 1981. Materials for a flora of the Tamilnadu Carnatic, Rapinat herbarium, India.

Matthew, K.M. 1991. An Excursion flora of Central Tamil Nadu, India. Oxford and IBH Publishing Co., New Delhi.

Nair, N.C. and Henry, A.N. 1983. Flora of TamilNadu, Series–I, Vol. I, *Botanical survey of India*, Southern – Circle, Coimbatore.

Rajendran, S.M. Chandrasekar, K. and Sundaresan, V. 2001. Ethnomedicinal lore of Seithur Hills—Southern Western Ghats, Tamil Nadu. *Ethnobotany.* 13: 101–109.

Sharma, P.P. and N.P. Singh, 2001. Ethnomedicinal uses of some edible parts in Dadra, Nagar Haveli and Daman (U.T.). *Ethnobotany*. 13: 121–125.

Venkatesan, G. Ganesan, S. Kesavan, L. and Banumathy, N. 2005. Ethnobotany of Ethnic group Thottianaickan of Tiruchirapalli District, Tamilnadu, India. In: *Advances in Medicinal Plants*, I: 59–75, Asian Medicinal Plants and Health Care Trust, Jodhpur.

11

in vitro Conservation Technique Through Indirect Somatic Embryogenesis from Seeds of *Atalantia monophylla* (L.) DC.

Arun, K. Das and P.S. Swamy

1. INTRODUCTION

Atalantia monophylla (L.) DC. is a small tree belonging to the family Rutaceae; resembles Citrus in general aspects and is considered a close wild relative of Citrus. The species is commonly called as wild lime, in Malayalam it is called as Kattunaragam and in Tamil Katuuelumichai. *A. monophylla* is used in Ayurveda, Folk and Siddha systems of medicine. It is used for the treatment of rheumatoid pain and glandular swelling (Sankaranarayan *et al.*, 2010). The root is believed to be antispasmodic (Kiritikar and Basu, 1999), the leaves are often applied for itching and other skin complaints (Panda, 2004) and acridone alkaloids isolated from roots were reported to have anti-allergic activity (Chukaew *et al.,* 2008). The chloroform extract of leaves has shown antiviral activity (Chansakaow *et al.*, 1996). Besides its medicinal value *A. monophylla* leaves are used as flavoring agent, berries are used for making pickles, juice of berries are used for dyeing purpose and wood as timber. It is also useful as a rootstock for breeding new cultivars of *Citrus* Linn. (Guha Bakshi *et al.*, 1999). This species has high medicinal value and the natural population is under threat due to habitat destruction, use of wood as firewood, collection of fruits for making pickles and harvesting of leaves, roots for medicinal purposes by the local communities. Hence there is a need for developing an appropriate conservation strategy for this species.

Somatic embryogenesis may be reproduced artificially by the manipulation of tissues and cells *in vitro*. Somatic embryos are derived from a single somatic cell or a group of somatic cells and they are morphologically similar to zygotic embryos. They are bipolar and bear typical embryonic organs including the radicle, hypocotyl and cotyledons. The greatest importance of somatic embryos is its practical application in large scale vegetative propagation (Arnold *et al.*, 2002). Also the somatic embryos obtained can be cryopreserved and can be used for long term storage and conservation. Somatic embryos can differentiate either directly from the explant without an intervening callus phase or indirectly after a callus phase (Williams and Maheswaran, 1986). The process of somatic embryogenesis is a suitable method of micropropagation and has the potential for mass propagation commercially at low cost per unit (Miah *et al.*, 2002). The somatic embryogenesis depends on several factors including genotype, explant source and culture medium (Lpoz-perez *et al.*, 2005). Therefore standardization of media and culture conditions for each and every species is necessary. The aim of the present study is to develop an *in vitro* conservation strategy through indirect somatic embryogenesis for this important medicinal plant.

2. MATERIALS AND METHODS

Mature fruits of *A. monophylla* were collected from Nagamalai hills, Madurai and the seeds were used as the primary explant. The fruits were washed thoroughly in running tap water for 10 minutes and placed in a detergent solution (5% Teepol) for 10 minutes. Then they are rinsed with distilled water for 5 minutes. Then the explants were taken to Laminar air flow and transferred to sterile conical flask and surface sterilized by first soaking in 5% sodium hypochlorite for 2-3 minutes, then in 80% ethanol for 30 seconds and finally soaked in 0.1% mercuric chloride solution for 5 minutes and finally rinsed with sterile distilled water for 5 times. The fruits are then cut open using sterile surgical blade (Lister No.10) and the seeds were then inoculated into MS medium supplemented with different concentrations of 2, 4-D. After the callus formation the callus obtained on the surface of the seeds were then sub-cultured into Basal MS media containing different concentrations of activated charcoal for indirect somatic embryogenesis. The cultures were maintained under fluorescent light at an intensity of 2000lux provided from fluorescent lamps with 16 hours of light and 8 hours of dark conditions. The obtained somatic embryos were further sub-cultured into basal MS media without the addition of activated charcoal to obtain individual plants. All the cultures were maintained in the culture room at temperature 25±2°C and relative humidity of 80-85%.

3. RESULTS AND DISCUSSION

In the present investigation a suitable protocol for indirect somatic embryogenesis of *A. monophylla* was developed. The mature seeds were used as the initial explant.

Table 1: Callus induction using MS media supplemented with different concentrations of 2, 4-D.

Sl.No.	Media	Percentage response (%)	Callus/Germination	Nature of callus
1.	Basal MS	100	Seed Germination	-
2.	MS+0.5mg/l 2,4-D	94.4	Seed Germination	-
3.	MS+1mg/l 2,4-D	66.6	Seed Germination	-
4.	MS+2mg/l 2,4-D	80.5	Callus	Yellow compact callus
5.	MS+3mg/l 2,4-D	88.8	Callus	Friable yellow callus
6.	MS+4mg/l 2,4-D	91.6	Callus	Friable light green callus

The surface sterilized explant was inoculated into MS media supplemented with different concentrations of 2, 4-D for callus induction. The best response was obtained when MS media supplemented with 4 mg/l 2, 4-D. Light green friable callus was obtained around the seed opening (Tab. 1). The obtained callus was then sub-cultured into basal MS media containing activated charcoal. Somatic embryogenesis was observed when the activated charcoal concentration was 2g/l (Fig. 2). In an earlier work in *Citrus macroptera* Mont. var. *anammensis* somatic embryogenesis was observed at 3g/l activated charcoal supplemented basal MS media (Miah *et al*, 2002), while in *C. sinensis* and *C. reticulata* the somatic embryos were obtained at 0.5g/l (Ricci *et al*, 2002). When the charcoal concentration is decreased callus proliferation takes place and when the concentration was increased browning of callus was observed. Mature somatic embryos were obtained after 3 weeks. After maturation of the somatic embryos, individual embryos were separated and sub-cultured into basal MS media without the addition of activated charcoal. Well developed plantlets were obtained after approximately 30 days (Fig. 3, 4). Using this technique 100% of the obtained somatic embryos converted to form healthy plantlets. The procedure requires approximately 12 weeks for obtaining individual plantlets. Genetic uniformity of the *in vitro* raised plantlets needs further investigation.

Activated charcoal is often used in plant tissue culture for its promotion of somatic embryogenesis. It is also known for its potent adsorption of growth inhibitors secreted by the tissues, such as phenolic compounds, sorption of agar impurities, media darkening; pH effects; and catalyzed breakdown of sucrose during sterilization with subsequent sorption of hydroxy methyl furfural. (Buffard-morel *et al.*, 1995; Moshkov *et al.*, 2008; Van winkle and Pulman, 1995). The benefits of using activated carbon include greater plant survival, greater plant growth, and improved plant quality and vigor. Somatic embryogenesis is one of the main micropropagation technique used for regeneration of Rutaceae members. The study shows that this technique can be used efficiently for conservation of *A. monophylla*. Genetic uniformity of the *in vitro* raised plantlets needs further investigation.

Fig. 1: Habit of *Atalanita monophylla* (L.) DC.

Fig. 2: Somatic embryos

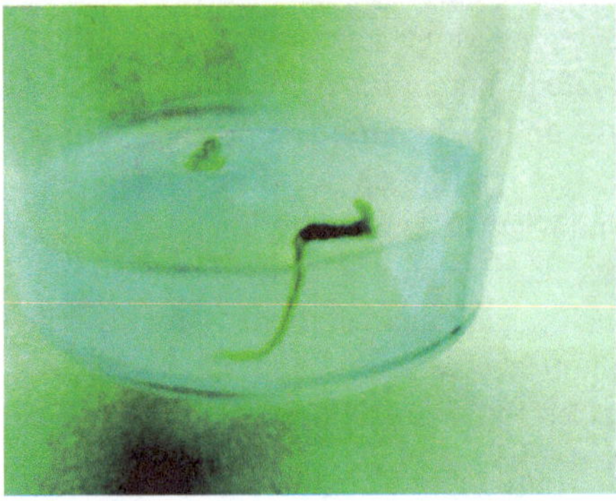

Fig. 3: Somatic embryo germination

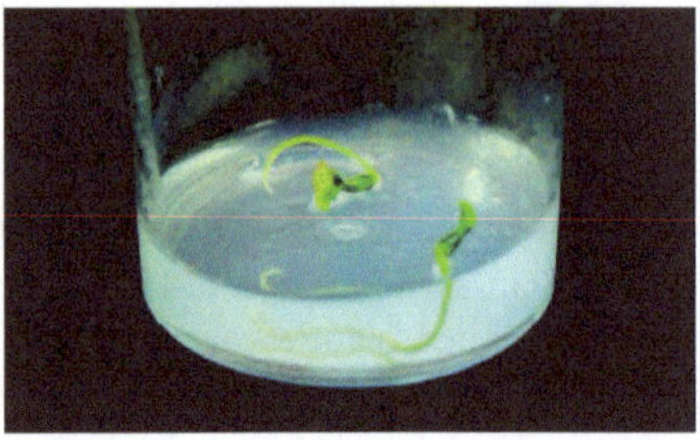

Fig. 4: Fully developed plantlets with shoot and root

REFERENCES

Arnold, S.V., Sabala, I., Bozhkov, P., Dyachok, J., Filonova, L., 2002. Developmental pathways of somatic embryogenesis, *Plant Cell Tiss Org*, 69, 233–249.

Buf'fard-morel, J., Verdeil, J.L., Dussert, S., Magnaval, C., Huet, C., Grosdemange, F., 1995. Initiation of somatic embryogenesis in coconut *(Cocos nucifera* L.). *Research and Practical Aspects,* 217-223.

Chansakaow, S., Ruangrungsi, N., Ishikawa, T., 1996. Isolation of pyropheophorbide A from the leaves of *Atalantia monophylla* (Rutaceae) as a possible antiviral active principle against herpes simplex virus type 2. *Chem Pharm Bull.*, 44, 1415-1417.

Chukaew, A., Ponglimanont, C., Karalai, C., Tewtrakul, S., 2008. Potential anti-allergic acridone alkaloids from the roots of *Atalantia monophylla*. *Phytochemistry*, 69, 2616–2620.

Guha Bakshi, D.N., Sensarma, P., Pal, D.C., 1999. A lexicon of medicinal plants in India. Vol. 1, Naya Prokosh Publications, Culcutta. p. 212.

Kirtikar, K.R., Basu, B.D., 1999. Indian Medicinal Plants. Bishen Singh Mahendra Pal Singh Publication, Dehradun. pp. 1655-1656.

López-pérez, A.J., Carreño, J., Martínez-cutillas, A., Dabauza, M., 2005. High embryogenic ability and plant regeneration of table grapevine cultivars (*Vitis vinifera* L.) induced by activated charcoal, *Vitis*, 44:79-85.

Miah, M.N., Islam, S., Hadiuzzaman, S., 2002. Regeneration of Plantlets through Somatic Embryogenesis from Nucellus Tissue of *Citrus macroptera* Mont. var. *anammensis* ('Sat Kara'). *Plant Tissue Cult.* 12:167-172.

Moshkov, I.E., Novikova, G.V., Hall, M.A., George, E.F. Plant growth regulators III: gibberellins, ethylene, abscisic acid, their analogues and inhibitors; miscellaneous compounds. Plant propagation by tissue culture, Springer Publication, Dordrecht, Netherlands, pp. 227-282.

Panda, H., 23004. Handbook on Medicinal Herbs with Uses. Asia Pacific Business Press, Inc., Delhi. pp. 166-167.

Sankaranarayanan, S., Bama, P., Ramachandran, J., Kalaichelvan, P.T., Deccaraman, M., Vijayalakshimi, M., Dhamotharan, R., Dananjeyan, B., Sathyabama, S., 2010. Ethnobotanical study of medicinal plants used by traditional users in Villupuram district of Tamil Nadu, India. *J. Med. Plants Res.*, 4:1089-1101.

Van-Winkle, S.C., Pullman, G.S., 1995. The Role of Activated Carbon in Tissue Culture Medium, *Energia*, 6,.

Williams, E.G., Maheswaran, G., 1986. Somatic embryogenesis: factors influencing coordinated behavior of cells as an embryogenic group. *Ann. Bot*, 57:443-462.

12

Pharmacognostic and Phytochemical Analysis of *Solanum anguivi* Lam. and *Solanum torvum* Sw.

Janakiraman, N. and Johnson, M.

1. INTRODUCTION

The therapeutic efficacy of many indigenous plants for various diseases has been described by traditional herbal medicine practitioners (Bharat and Parabia, 2010). Natural products are the source of synthetic and traditional herbal medicine. They are still the primary health care system in some parts of the world (Mahmud *et al.*, 2010). In India, local empirical knowledge about medicinal properties of plants is the basis for their uses as a home remedies. It is generally accepted by many Indians and elsewhere in the world that beneficial medicinal effects can be obtained by ingesting plant products. Plants bear the basis of many traditional medicines throughout the world for thousands of years and continue to provide new remedies to mankind (Patel, 2009). The World Health Organization (WHO) also encourages, recommends and promotes the use of traditional and herbal remedies in the National Health Care Program because these medicines are easily available at low cost, comparatively safe and people have faith in such remedies (Wani, 2007). Herbal medicines are promising choice over modern synthetic drugs. Due to the serious obsession with the modern medicinal system, people have started looking at the ancient healing systems like Ayurveda, Siddha and Unani (Shankar and Ved, 2003). However a key obstacle, which has hindered the acceptance of the herbal medicines, is the lack of documentation and quality control. Hence, there is a need for documentation of research work and effort towards standardization of the plant materials to be used as medicine (Dahanukar *et al.*, 2000). The process of standardization can be achieved by stepwise pharmacognostic studies (Calixto, 2000) including identification and authentication of the plant material. Correct identification and quality assurance of the starting materials is an essential prerequisite to ensure reproducible quality of herbal medicine which will contribute to its safety and efficacy. Simple pharmacognostic techniques used in standardization of plant material include its morphological, anatomical and biochemical characteristics.

Solanum is the most representative genus of the family Solanaceae with about 1400 species and cosmopolitan distribution (Bohs, 2005). *Solanum anguivi* Lam. (Vine of Sodom) is a rare ethnomedicinal herb which is used as therapeutic agent for various diseases. The roots are carminative and expectorant useful in treating coughs, cultarrhal affections, dysuria, colic, nasal ulcers, ingredient of dasamula, asthma, difficult parturition, tooth ache, cardiac disorder, worm complaints, spinal cord disorder, nervous disorder and fever. The leaves and fruits rubbed up with sugar are used as external application for itch (Yoganarasimhan, 1996). The stem, fruits, roots, flowers and leaves contain glycoalkaloids, steroidal alkaloids, solamargine and solasoline (Chopra *et al.*, 1994). *Solanum torvum* Sw. (Solanaceae), commonly known as Turkey berry is an erect spiny shrub native to Africa and West Indies (Adjanohoun *et al.*, 1996). The fruits and leaves are widely used in Camerooninan folk medicine. The plant is cultivated in the tropics for its sharp tasting immature fruits. It is used in the treatment of stomach pain and skin infections (Siemonsma and Piluek, 1994). It possesses antimicrobial (Ajaiyeoba, 1999; Chah *et al.*, 2000) antiviral (Arthan *et al.*, 2002), imuno-secretory (Israf *et al.*, 2004), antiulcer (Nguelefack *et al.*, 2008), antioxidant (Sivapriya and Srinivas, 2007), analgesic and anti-inflammatory (Ndebia *et al.*, 2007), cardiovascular and anti-platelet aggregation activities (Nguelefack *et al.*, 2008). *S. torvum* contains alkaloids like jurubine, paniculogenin, solasodine, solasodien, torvogenin, steroids like sisalogenin, torvogenin and sisalogenone, flavonoids such as torvanol A and torvoside H, and sapogenin. It also contains a number of potentially pharmacologically active chemicals like isoflavonoid sulfate and steroidal glycosides, chlorogenone and neochlorogenone, triacontane derivatives, 22-α-O-spirostanol oligoglycosides and 26-O-α-glucosidase (Iida *et al.*, 2005; Arthan *et al.*, 2006). With this information, the present study was aimed to reveal the pharmacognostic and phytochemical profile of *S. anguivi* and *S. torvum*.

2. MATERIALS AND METHODS

2.1 Collection and processing of plant materials

The whole plants of *Solanum anguivi* Lam. were collected from Courtallam and *Solanum torvum* Sw. from Palayamkottai, Tirunelveli, South India. The collected plant materials were washed under running tap water to remove the surface pollutants and rinsed twice with distilled water. The plants were cut into small pieces and shade dried at room temperature for 20 days. The shade dried materials were powdered using the homogenizer and used for further studies.

2.2 Extraction of plant material

Shade dried and powdered plant materials were packed in small thimbles separately and successively extracted with various solvents viz., petroleum ether, benzene, chloroform, ethanol and aqueous using soxhlet apparatus for 8 h continuously. The different solvent extracts were concentrated by rotary vacuum evaporator.

2.3 Physico-chemical parameters and fluorescence analysis

Various ash types and extractive values of different extracts of *S. anguivi* and *S. torvum* were determined by following standard method (Indian Pharmacopoeia, 1996). Fluorescence analysis was carried out as per the standard procedures (Lala, 1993). In the present study, the whole plant powder was treated with 1N aqueous NaOH, 1N ethanolic NaOH, acids like 1N HCl, 1N H_2SO_4, HNO_3 and various extracts like petroleum ether, benzene, chloroform, ethanol and aqueous. These were subjected to fluorescence analysis in visible/daylight and UV light (365 nm).

2.4 Phytochemical analysis

The different extracts of *S. anguivi* and *S. torvum* were subjected to qualitative tests for the identification of various secondary metabolites as per the standard procedure (Harborne, 1998).

3. RESULTS AND DISCUSSION

The morphological variations found in both the species of *Solanum* are illustrated in Table 1. The ash values and extractive values of *S. anguivi* and *S. torvum* were represented in Table 2. The moisture content of *S. anguivi* was found to be 67.34% whereas in *S. torvum* it was about 70.39%. The aerial parts of *S. anguivi* possess 10.1% of total ash which was almost nearer to *S. torvum* (12.8%). The water soluble ash content of *S. torvum* and *S. anguivi* was 47.7% and 43.6% respectively. The acid insoluble ash of *S. anguivi* (0.8%) is almost close to *S. torvum* (0.7%). *S. torvum* showed more extractive values than *S. anguivi*. The ethanolic extract showed the highest extractive value of 1.22% and petroleum ether extract showed the lowest value of 0.02%. Fluorescence analysis of the aerial part powders of *S. anguivi* and *S. torvum* treated with different chemical reagents and the extracts of powder in various solvents can be used as a diagnostic tool for identifying the adulterant from the source. The results of fluorescence analysis examined under ordinary and UV light (365 nm) were presented in Table 3.

Table 1: Morphological variations in *S. anguivi* and *S. torvum*

Name of the species	Size(m)	Habit	Leaf characters	Fruit
S. anguivi	1.8	Shrub or under shrub	Sub entire, ovate, sinuate or lobed, stellately woolly beneath	Berries globose reddish or dark yellow when young
S. torvum	4.0	Shrub	Soft hairy above, stellately tomentose below	Berries globose, smooth green

Table 2: Physico-chemical characters of aerial parts of *S. anguivi* and *S. torvum*

	Ash values	
Particulars	*S. anguivi*	*S. torvum*
Loss of weight on drying	67.34	70.39
Total ash	10.1	12.8
Acid insoluble ash	0.8	0.7
Water soluble ash	43.6	47.7
Extractive values		
Extracts	*S. anguivi*	*S. torvum*
Petroleum ether	0.16	0.02
Benzene	0.53	0.18
Chloroform	0.16	0.36
Ethanol	0.31	1.22
Aqueous	0.11	0.52

Table 3: Fluorescence characters of *S. anguivi* and *S. torvum*

Particulars of treatment	*S. anguivi*		*S. torvum*	
	Ordinary light	UV light (365 nm)	Ordinary light	UV light (365 nm)
Dry powder as such	Green	Green	Brownish green	Green
Powder + NaOH (aqueous)	Greenish yellow	Green	Greenish brown	Greenish black
Powder + NaOH (ethanolic)	Brownish green	Dark green	Brownish green	Greenish black
Powder + 1N HCl	Yellowish green	Green	Brownish green	Green
Powder + 1N H_2SO_4 (1:1)	Yellowish green	Light green	Brownish green	Blackish green
Powder + HNO_3 (1:1)	Brown	Dark green	Brown	Brownish green
Petroleum ether	Brownish green	Dark green	Brownish green	Greenish black
Benzene	Greenish brown	Blackish green	Greenish brown	Greenish black
Chloroform	Yellowish green	Blackish green	Yellowish green	Greenish black
Ethanol	Dark green	Blackish green	Brownish green	Blackish green
Aqueous	Brownish green	Dark green	Brownish green	Dark green

Standardization is essential measure for quality, purity and sample identification in the pharmaceutical industries. Studies on physico-chemical characters serve as valuable source of information and provide suitable standards to determine the quality of plants. The authenticity and purity of plant samples are determined by their ash values which act as important quantitative standards and also helps in detecting the presence or absence of any foreign organic matter such as metallic salts and siliceous contamination. Due to the possibility of the presence of non-physiological substances, the total ash value of a crude plant sample is always accompanied with the determination of water soluble and insoluble ash (Musa *et al.*, 2006). Extractive values are useful for the determination of adulterated drugs (Nayak and Patel, 2010). The fluorescent method is sufficiently sensitive and enables the precise and accurate determination of the analyte over a satisfactory concentration range without several time consuming dilution steps prior to analysis of pharmaceutical samples. Correct identification and quality assurance of the starting materials is an essential prerequisite to ensure reproducible quality of herbal medicine which will contribute to its safety and efficacy (Pimenta *et al.*, 2006). Hence, the results of the present pharmacognostical characters may be helpful to identify the standard medicinal plant for the commercial application.

The preliminary phytochemical screening of ethanolic, benzene, chloroform, petroleum ether and aqueous extracts of *S. anguivi* and *S. torvum* were tabulated in Table 4 and 5. Steroids, sugars and phenolic compounds showed their presence in all the tested five extracts of *S. anguivi.* Alkaloids and flavonoids were present in all the extracts except chloroform, tannins were present in all the extracts except aqueous, triterpenoids, catechin and saponins showed their presence only in petroleum ether extract of *S. anguivi*. In *S. torvum,* steroids, sugars and reducing sugars were present in all the tested three extracts where as tannins illustrated their existence in all the extracts except aqueous extract. Phenolics and flavonoids demonstrated their occurrence in three different extracts whereas alkaloids showed their presence in two different extracts of *S. torvum*. Anthraquinone failed to show their presence in all the tested extracts of *S. anguivi* and *S. torvum*.

Plants can produce different types of secondary metabolites which have been subsequently utilized by humans for their valuable characters in a diverse array of applications (Zwenger and Basu, 2008). Secondary metabolites include compounds produced in response to stress, such as the case when acting as a deterrent against herbivores (Keeling and Bohlmann, 2006). Currently, there is an increased interest in natural substances with valuable medicinal compounds. Plant terpenoids are used extensively for their aromatic qualities. They played a major role in traditional herbal remedies and are used for antibacterial, antineoplastic and other pharmaceutical functions (URL, 2012). Flavonoids are ubiquitous in photosynthesising cells and occur widely in the plant kingdom. They are found in fruit, vegetables, nuts, seeds, stems and owers and represent a common constituent of the human diet (Deshmukh *et al.*, 2008). Flavonoids are known to inhibit bacterial growth (Mbuh *et al.*, 2007). Tannins are known to possess antimicrobial and antioxidant activities (Rievere *et al.*, 2009). They also have potential values as cytotoxic and antineoplastic agents (Aguinaldo *et al.*, 2005). Saponins are mild

Table 4: Preliminary phytochemical analysis of *S. anguivi*

Name of the extract	Steroids	Triterpenoids	Reducing sugars	Sugars	Alkaloids	Phenolics	Flavonoids	Catechin	Saponins	Tannins	Anthraquinone
Ethanol	+	-	-	+	+	+	+	-	-	+	-
Benzene	+	-	+	+	+	+	+	-	-	+	-
Chloroform	+	-	+	+	-	+	-	-	-	+	-
Petroleum ether	+	+	+	+	+	+	+	+	+	+	-
Aqueous	+	-	+	+	+	+	+	-	-	-	-

Table 5: Preliminary phytochemical analysis of *S. torvum*

Name of the extract	Steroids	Triterpenoids	Reducing sugars	Sugars	Alkaloids	Phenolics	Flavonoids	Catechin	Saponins	Tannins	Anthraquinone
Ethanol	+	-	+	+	+	-	+	-	-	+	-
Benzene	+	-	+	+	+	-	+	-	-	+	-
Chloroform	+	-	+	+	-	+	-	-	-	+	-
Petroleum ether	+	+	+	+	-	+	-	-	+	+	-
Aqueous	+	-	+	+	-	+	+	-	-	-	-

detergents used in intracellular histochemistry staining to allow antibody access to intracellular proteins. It is known to have antioxidant, anticancer, anti-inflammatory and anti-fungal properties (De-Lucca *et al.*, 2005). Saponins have been implicated as a bioactive antibacterial agent of plants (Mandal *et al.*, 2005; Manjunatha, 2006). Alkaloids are used in medicine in the form of salts including anti-arrhythmic, anti-cholinergic, anti-tumor, vasodilating, anti-hypertensive, cough, anesthetic, anti-microbial and anti-protozoal agents (URL, 2012a). Phenolic compounds exhibit a wide range of physiological properties such as anti-inflammatory, anti-microbial, anti-allergenic, antioxidant, cardioprotective and vasodilatory effects (Manach and Mazur, 2005). The results of the present study validated the existence of alkaloids, flavonoids, saponins, tannins, phenols, terpenoids etc. in different extracts of both the species of *Solanum* which can be used as anti-arrhythmic, anti-cholinergic, anti-tumor, vasodilating, anti-hypertensive, anesthetic, antineoplastic, antioxidant, anti-inflammatory, anti-microbial, anti-allergenic and anticancer agents in the near future.

4. CONCLUSION

The reported pharmacognostical characters can serve as a valuable source of information and provide a suitable diagnostic tool for identification of adulterants as well as standardization of herbal drugs of folk medicinal practice of present era and enrichment of Ayurvedic pharmacopoeia. It will be very much useful for the preparation of a monograph of the plant and also be immense for carrying out further research and revalidation of its use.

REFERENCES

Adjanohoun, J.E., Aboubakar, N., Dramane, K., Ebot, M.E., Ekpere, J.A., Enoworock, E.G., Foncho, D., Gbile, Z.O., Kamanyi, A., Kamoukom, J., Keeta, A., Mbenkum, T., Mbi, C.M., Mbielle, A.L., Mbome, I.L., Mubiru, N.K., Naney, W.L., Nkongmeneck, B., Satabie, B., Sofowa, A., Tanze, V., Wirmum, C.K., 1996. Traditional medicine and pharmacopeia contribution to ethnobotanical and floristic studies in Cameroon. In: CNPMS, Porto-Novo, Benin, 50-52.

Aguinaldo, A.M., Espeso, E.I., Guevara, B.Q., Nonato, M.G., 2005. Phytochemistry. In: Guevara, B.Q. (Ed.), A Guidebook to Plant Screening: Phytochemical and Biological. University of Santo Tomas, Manila, Philippines, 121-125.

Ajaiyeoba, E.O., 1999. Comparative phytochemical and antimicrobial studies of *Solanum macrocarpum* and *Solanum torvum* leaves. *Fitoterapia* , 70:184-186.

Arthan, D., Kittakoop, P., Esen, A., Svasti, J., 2006. Furostanol glycoside 26- O-â glucosidase from the leaves of *Solanum torvum. Phytochemistry* , 67: 27-33.

Arthan, D., Svasti, J., Kittakoop, P., Pittayakhachonwut, D., Tanticharoen, M., Thebtaranonth, Y., 2002. Antiviral isoflavonoid sulfate and steroidal glycosides from the fruits of *Solanum torvum. Phytochemistry,* 59: 459-463.

Bharat, G, Parabia, M.H., 2010. Pharmacognostic Evaluation of Bark and Seeds of *Mimusops elengi* L. *International Journal of Pharmacy and Pharmaceutical Sciences.* 2(4): 110-113.

Bohs, L., 2005. Major clades in *Solanum* based on ndhF sequences. In: Keating, R.C., Hollowell, V.C., (Eds.), Croat monographs in Systematic Botany from the Missouri Botanical Garden. A festschrift for William G. D'Arcy: the legacy of a taxonomist. St. Louis: Missouri Bot Garden Press. 104: 27-49.

Calixto, J.B., 2000. Efficacy, safety, quality control, marketing and regulatory guidelines for herbal medicines (Phytotherapeutic agents). *Braz. J. Med. Biol. Res*., 33(2): 179-189.

Chah, K.F., Muko, K.N., Oboegbulem, S.I., 2000. Antimicrobial activity of methanolic extract of *Solanum torvum* fruit. *Fitoterapia*. 71: 187-189.

Dahanukar, S.A., Kulkarni, R.A., Rege, N.N., 2000; *Pharmacology* of medicinal plants and natural products. *Ind. J Pharmacol*. 32: 81-118.

De-Lucca, A., Cleveland, T., Rajasekara, K., Boue, S., Brown, R., 2005. Fungal properties of CAY-1, a plant saponin, for emerging fungal pathogens 45th Interscience Conference in Antimicrobial Agents and Chemotherapy. F-490, 180.

Deshmukh, T., Yadav, B.V., Badole, S.L., Bodhankar, S.L., Dhaneshwar, S.R., 2008; Antihyperglycaemic activity of alcoholic extract of *Aerva lanata* (L.) A. L. Juss. Ex J. A. Schultes leaves in alloxan induced diabetic mice. *J Appl Biomed*. 6: 81-87.

Harborne, J.B., 1998. Phytochemical Methods, A guide to modern techniques of plant analysis, 3rd Edn. Chapman and Hall, New York, 1-150.

Iida, Y., Yanai, Y., Ono, M., Ikeda, T., Nohara, T., 2005. Three unusual 22-β- O-23-hydroxy-(5α)-spirostanol glycosides from the fruits of *Solanum torvum*. *Chemical & Pharmaceutical Bulletin*. 53: 1122-1125.

Indian Pharmacopoeia, 1996, Controller of Publication, Government of India.

Israf, D.A., Lajis, N.H., Somchit, M.N., Sulaiman, M.R., 2004. Enhancement of ovalbumin-specific IgA responses via oral boosting with antigen coadministered with an aqueous *Solanum torvum* extract. *Life Sci*.75: 397-406.

Keeling, C.I., Bohlmann, J., 2006. Genes, enzymes, and chemicals of terpenoid diversity in the constitutive and induced defence of conifers against insects and pathogens. *New Phytol*. 170: 657-675.

Lala, P.K., 1993. Lab Manuals of Pharmacognosy. CSI Publishers and Distributors, 5th Edition, Calcutta.

Mahmud, S., Shareef, H., Ahmad, M., Gouhar, S., Rrizwani, G.H., 2010. Pharmacognostic studies on fresh mature leaves of *Holoptelea integrifolia* (Roxb) Planch. *Pak J Bot*. 42(6): 3705-3708.

Manach, Mazur, 2005. Bioavailability and bio-efficacy of polyphenols in humans. *Am. J. Clin. Nutr*. 81: 230S-242S

Mandal, P., Sinha Babu, S.P., Mandal, N.C., 2005. Antimicrobial activity of Saponins from *Acacia auriculiformis*. *Fitoterapia*. 76(5): 462-565.

Manjunatha, B.K., 2006. Antibacterial activity of *Pterocarpus santalinus*. *Ind. J. Pharm. Sci*. 68(1): 115-116.

Mbuh, F.A., Asika, I.S., Doughari, J.H., 2007. Studies on antibacterial activity of leaf extracts of *Psidium guajava* L. Biological and Environmental Sciences Journal for the Tropics 5(1), 44-47.

Musa, K.Y., Katsayal, A.U., Ahmed, A., Mohammed, Z., Danmalam, U.H., 2006. Pharmacognostic investigation of the leaves of *Gisekia pharmacioides*. *African Journal of Biotechnology*. 5: 956-957.

Nayak, B.S., Patel, K.N., 2010. Pharmacognostic studies of the *Jatropha curcas* leaves. *International Journal of PharmTech Research*. 2(1): 140-143.

Ndebia, E.J., Kamga, R., Nchunga-Anye, N.B., 2007. Analgesic and anti-inflammatory properties of aqueous extract from leaves of *Solanum torvum* (Solanaceae). *AJTCAM*. 4: 240-244.

Nguelefack, T.B., Feumebo, C.B., Watcho, G.A.P., Tatsimo, S., Atsamo, A.D., Tane, P., Kamanyi, A., 2008. Anti-ulcerogenic properties of the aqueous and methanol extracts from the leaves of *Solanum torvum* Swartz (Solanaceae) in rats. *J. Ethnopharmacol.* 119 (1):135-140.

Nguelefack, T.B., Mekhfi, H., Dimo, T., Afkir, S., Nguelefack-Mbuyo, E.P., Legssyer, A., Ziyyat, A., 2008. Cardiovascular and anti-platelet aggregation activities of extracts from *Solanum torvum* (Solanaceae) fruits in rat. *J.Com. Int. Med.* 5(1):1-11.

Patel, S.S., 2009. Morphology and Pharmacology of *Passiflora edulis*. *Journal of Herbal medicine and Toxicology*. 3(1): 1-6.

Pimenta, A.M., Montenegro, M.C., Araujo, A.N., Martine, J.C., 2006. Application of sequential injection analysis to pharmaceutical analysis. *Journal of Phar. Biomed. Annuls.*, 40:16-34.

Rievere, C., Van Nguyen, J.H., Pieters, L., Dejaegher, B., Heyden, Y.V., Minh, C.V., Quetin-Leclercq, J., 2009. Polyphenols isolated from antiradical extracts of *Mallotus metcalfianus*. *Phytochemistry*. 70: 86-94.

Chopra, R.N., Chopra, I.C., Handi, K.L., Kapor, L.D., 1994. Indigenous drugs of India II edition Academic publishers, Calcutta.

Shankar, D., Ved, D.K., 2003. A balanced perspective for management of Indian medicinal plants. *Indian Forester*. 129: 275-288.

Siemonsma, J.S., Piluek, K., 1994. Plant resources of South-East Asia 8 (PROSEA). Indonesia: Bogor.

Sivapriya, M., Srinivas, L., 2007. Isolation and purification of a novel antioxidant from the water extract of Sundakai (*Solanum torvum*) seeds. *Food Chemistry*. 104: 510-517.

URL, 2012. http://en.wikipedia.org/wiki/Terpenoid

URL, 2012a. http://en.wikipedia.org/wiki/Alkaloids

Wani, M.S., 2007. Herbal medicine and its standardization. Pharmaceutical Reviews 5(6).

Yoganarasimhan, S.N., 1996. Medicinal plant of India Vol. I Interline Publishing Pvt. Ltd. Bangalore, India.

Zwenger, S., Basu, C., 2008. Plant terpenoids: applications and future potentials. *Biotechnol Mol Biol Rev*. 3(1): 1-7.

13

Antioxidant Activity of *Ophiorrhiza pectinata* Arn. (Rubiaceae)

G. M. Lekshmi and A. Gangaprasad

1. INTRODUCTION

Antioxidants are defined as molecules which can be safely interact with free radicals and terminate the chain reaction before vital molecules are damaged (Nidhi and Gaurav, 2012). Naturally occurring antioxidants to replace synthetic antioxidants which are being restricted due to their carcinogenicity (Abdel Hameed, 2009). The antioxidant phytochemicals such as polyphenols, flavanoids and related compounds found in medicinal plants have received increasing attention for their potential role in prevention of human diseases (Upadhyay *et al.*, 2010).

The genus *Ophiorrhiza* belongs to the family Rubiaceae and consists of about 150 species and distributed in tropical and subtropical Asia. Species of *Ophiorrhiza* are economically important and many of these species contain a high value anticancerous compound, a cytotoxic alkaloid camptothecin and other valuable alkaloids. *Ophiorrhiza pectinata* Arn. is a herbaceous plant and is distributed over peninsular India and Srilanka. A preliminary study conducted in *O. pectinata* also contains this important alkaloid (0.0040mg/g dry wt). Leaves and stem contain traces of hydrocyanic acid. The main objective of the present study is to evaluate the antioxidant and radical scavenging property of the chloroform methanol and water extract of whole plant of *O. pectinata*.

2. MATERIALS AND METHODS

Samples of *O. pectinata* were collected from Ponmudi in the Western Ghats forests of Kerala and were maintained in the greenhouse of the Department of Botany, University of Kerala, Kariyavattom. The plant materials were shade dried and then powdered. The material was extracted using Soxhlet apparatus and concentrated in a rotary evaporator (Super fit Rotavap, India). All the chemicals used were of analytical grade.

Total phenols was determined by using Folin-ciocalteus reagent and external calibration with chlorogenic acid (Slinkard and Singleton, 1997). Briefly, 0.5ml of the extract solution in acetone, 2.5 ml of Folin-ciocalteus reagent ($1/10^{th}$ dilution) was added and the contents mixed thoroughly. After 4 minutes, 2ml of 7.5% sodium carbonate was added, and then the mixture was incubated at 45^0 C for 15 minutes. The absorbance was measured at 765 nm. The total phenolic contents were expressed as chlorogenic acid equivalent in microgram per mg of the extract using a standard curve generated with chlorogenic acid.

The concentration of total flavonoids was determined by using the aluminium chloride colorimetric method from the procedure reported by Zhuang *et al.,* (1992). Briefly 0.2 ml of extract solution in a test tube was mixed with 0.1 ml of 10% aluminium chloride, 0.1ml of 1M Potassium acetate and 4.6 ml of distilled water. The absorbance was measured at 415 nm after incubation at room temperature for 45 minutes. The total flavonoid contents were expressed as quercetin equivalent in microgram per mg of the extracts using a standard curve generated with quercetin.

Reductive potential was carried out by the method of Oyaizu (1986). Different concentrations of the extracts (10-100 μg/ml) were prepared. To all the test tubes 2.5ml of sodium phosphate buffer (0.2M, pH 6.6) followed by 2.5 ml of 1% potassium ferricyanide solution was added. The contents were vortexed well and incubated at 50 ^{0}C for 20 minutes. After incubation, 2.5 ml of 10% trichloroacetic acid was added to all the tubes and centrifugation was carried out at 3000 rpm for 10 minutes. To 5 ml of the supernatent, 5ml of distilled water was added. To this about 1ml of 1% ferric chloride was added to each test tube and incubated at 35^0 C for 10 min. The absorbance was read at 700 nm. Higher absorbance indicated higher reducing power. Ascorbic acid was used as standard, all the tests were performed in triplicate.

DPPH radical scavenging activity was determined according to the method described by Blois (1958). Different concentrations of the whole plant extracts (10-100μg/ml) were prepared in 100% methanol and subjected to antioxidant tests. To the methanolic solution of DPPH (10μM), 0.05ml of test compounds of different concentrations (10 - 100μg/ml) dissolved in 100% methanol were added. An equal amount of methanol was added to the control. After 20 minutes the absorbance of the samples were read at 517 nm in a UV-Visible spectrophotometer (Shimadzu UV-1700, Japan) and percentage of inhibition was calculated.

% inhibition = [(Absorbance of control - Absorbance of sample) / Absorbance of control] x 100

3. RESULTS AND DISCUSSION

Total phenolic content in 100μg was estimated from three extracts (chloroform, methanol and water). Among these methanol extract have the highest phenolic content of 22 μg/mg followed by chloroform extract 20 μg/mg and then water extract having 19.5 μg/mg. Total flavonoid content in 100 μg/mg was estimated from chloroform, methanol and water extracts. Among these methanol extract have the highest flavonoid content of 19 μg/mg followed by chloroform extract 17.5 μg/mg and then water extract having 16 μg/mg.Chloroform and methanol extract showed potent antioxidant activity based on DPPH radical inhibition and reducing power tests in dose dependent manner (Table 1 & 2). Chloroform and methanol extract showed maximum inhibition of 62.6% and 58.2% at 100μg/ml respectively. Reductive potential of chloroform and methanol extract was 0.291 and 0.288 at 100μg/ml. The chloroform and methanol extracts exhibited the IC_{50} value of 65 μg /ml and 70 μg/ml

Table 1: Reductive potential of *O. pectinata* whole plant

Con (μg/ml)	Absorbance at 700nm inhibition		
	Chloroform extract (Mean+SE)	Methanol extract (Mean+SE)	Water extract (Mean+SE)
20	0.257 + 0.0008	0.25 + 0.0011	0.235 + 0.0015
40	0.262 + 0.0014	0.256 + 0.0008	0.242 + 0.002
60	0.269 + 0.0012	0.263 + 0.0003	0.253 + 0.002
80	0.278 + 0.0015	0.280 + 0.0008	0.258 + 0.0041
100	0.291 + 0.0016	0.288 + 0.0023	0.276 + 0.002

Reductive potential of ascorbic acid (60 μg/ml) is 0.893 + 0.0073

Table 2: Free radical scavenging activity of *O. pectinata* whole plant by DPPH radical inhibition

Con (μg/ml)	% of inhibition		
	Chloroform extract (Mean+SE)	Methanol extract (Mean+SE)	Water extract (Mean+SE)
20	26.8 + 0.0008	16.4 + 0.0008	2.98 + 0.0011
40	32.8 + 0.0023	32.8 + 0.0014	8.95 + 0.001
60	47.76 + 0.0011	46.2 + 0.0009	29.85 + 0.0014
80	59.70 + 0.0014	50.74 + 0.001	37.3 + 0.0014
100	62.6 + 0.0011	58.2 + 0.001	38.80 + 0.002

Radical scavenging activity of ascorbic acid (60 μg/ml) is 0.026 + 0.00003

Free radical reactions, especially with participation of oxidative radicals, have been shown to be involved in many biological processes that cause damage to lipids, proteins, membranes and nucleic acids, thus giving rise to a variety of diseases (Datta *et al.,* 2000). Antioxidants are important in disease prevention in both plants and animals, inhibiting or delaying the oxidation of biomolecules by preventing the initiation or propagation of oxidizing chain reactions. Synthetic antioxidants require extensive and expensive tests to ascertain their safety for food applications, and for this reason, there is interest in the use of naturally occurring antioxidants (Frankel,1995). Many natural products have been reported to contain large amounts of antioxidants other than vitamin C, E and carotenoids (Javanmardi *et al.,* 2003). Antioxidant activity may be mainly due to the presence of phenolic compounds such as flavonoids (Pietta, 1998), phenolic acids and phenolic diterpenes (Shahidi *et al.,* 1992). In *O. pectinata* phenol and flavonoid content was present higher in chloroform and methanol extract. Flavonoids and phenolic compounds are widely distributed in plants which have been reported to exert multiple biological effects, including antioxidant, free radical scavenging abilities, anti-inflammatory, anticarcinogenic etc (Miller, 1996). DPPH assay is one of the most widely used methods for screening the antioxidant activity of plant extracts. Antioxidants react with DPPH, which is a stable free radical, and convert it to 1, 1,- diphenyl-1-picryl hydrazine. The stable free radical DPPH has been widely used to test the free radical-scavenging the ability of various dietary antioxidants (Brand Williams *et al.,* 1995). In the present study the radical scavenging activity of the chloroform and methanol extracts of *O.pectinata* whole plant extract was excellent and IC_{50} of chloroform and methanol extract was 65μg and 70μg respectively. In *Salvia sclarea* also chloroform extract had stronger total antioxidant activity (Ilhami Golcin *et al.,* 2004). In the present study reductive potential was also higher in chloroform and methanol extract, it has been reported that reducing power was associated with antioxidant activity (Yen and Duh, 1993).

4. CONCLUSION

Methanol, water and chloroform extract of the whole plant of *Ophiorrhiza pectinata* were subjected to the estimation of total phenol & flavonoid content in 100 μg. Phenolic content was higher in methanol extract having 22 μg/ml followed by chloroform extract 20 μg/ml and then water extract 19.5 μg/ml. Flavonoid content was also higher in methanol extract having 19 μg/ml followed by chloroform extract 17.5 μg/ml and then water extract having 16 μg/ml, and the extracts were subjected to antioxidant activity based on DPPH radical inhibition and reducing power tests in dose dependent manner. Chloroform and methanol extract exhibited the IC_{50} value of 65μg/ml and 70μg/ml. Reductive potential of chloroform and methanol extract was 0.291 and 0.288 respectively at 100μg/ml.

REFERENCES

Abdel- Hameed, E.S., 2009. Total phenolic contents and free radical scavenging activity of certain Egyptian Ficus species leaf samples. *Food Chemistry* 114:1271-1277.

Blois, M.S., 1958. Antioxidant determinations by the use of a stable free radical. *Nature* 181, 1199-1200.

Brand-williams,W., Cuvelier, M.E., Berset, C., 1995. Use of a free radical method to evaluate antioxidant activity. *Lebensmittel-wissenschaft und-Technologie* 28:25-30.

Datta, K., Sinha, S., Chattopadhyayay, P., 2000. Reactive oxygen species in health and disease. *National Medical Journal of India* 13, 310.

Frankel, E.N., 1995. Natural and biological antioxidant systems. Their mechanisms of action, applications and implications. *Lipid Technology* 7:77-80.

Ilhami Golcin., Metin Tansu uguz., Munir oktay Sukru Beydemir., Irfan Kufrevioglu O., 2004. Evaluation of the antioxidant and antimicrobial activities of clary sage (*Salvia Sclarea* L.). *Turkish Journal of Agriculture and Forestry* 28:25-33.

Javanmardi, J., Stushnoff, C., Locke, E., and Vivanco, J. M., 2003. Antioxidant activity and total phenolic content of Iranian *Ocimum* accessions. *Food Chemistry* 83:547-550.

Miller, A.L., 1996. Antioxidant flavonoids : structure, function and clinical usage. *Alternative Medicine Review* 1:103-111.

Nidhi, S., Gaurav, S., 2012. Antioxidants from natural source: Ray of hope for oxidative damage. *Journal of Pharmaceutical and Scientific Innovation* 1:14-17.

Oyaizu, M., 1986. Studies on product of browning reaction prepared from glucose amine. *Japanese Journal of Nutrition* 44:307-315.

Shahidi, F., Wana sundara, P., 1992. Phenolic antioxidants. *Critical Reviews in Food Science and Nutrition* 32:67-103.

Slinkard, K., Singleton, V.L. 1977. Total phenol analysis. *Automation and comparison with manual methods. American Journal of Enology and Viticulture* 28:49-55.

Upadhyay, N.K., Yogendrakumar, M.S., Gupta, A., 2010. Antioxidant, cytoprotective and antibacterial effects of sea buckthorn (*Hippophae rhamnoides* L.). *Food and Chemical Toxicology* 48:3443-3448.

Yen, G.C., Duh, P.D., Tsai, C.L., 1993. Relationship between antioxidant activity and maturity of peanut hulls. *Journal of Agricultural and Food Chemistry* 41:67-70.

Zhuang, X.P., Lu, Y.Y.N.D., Yang, G.S., 1992. Extraction and determination of flavonoid in *ginkgo. Chinese Herbal Medicine* 23:122-124.

14

Antibacterial Activity of Clove bud (*Syzygium aromaticum* L.)

Meera Nandhini, E., Aruna Sharmili, Ranjitha, P. and Umamaheswari, K.

1. INTRODUCTION

The discovery and development of antibiotics have led to dramatic improvement in the ability to treat infectious diseases with emergence of drug – resistant organisms due to irrational and overuse of antibiotics, failure to complete a course of treatment, genetic versatility of microbes and horizontal transfer of resistant genes among bacterial species had led to antibiotic resistance (Amit and Shailendra, 2006; Aibinu, 2007).The need for alternative antimicrobial agents has renewed the interest on plants as a sources of antimicrobial agents due to their use historically and the fact that a good portion of the world's population, particularly in developing countries, rely on plants for the treatment of infectious and non-infectious diseases.

Syzygium aromaticum

Clove (*Syzygium aromaticum*) is the aromatic dried flower buds of a tree in the family Myrtaceae. The traditional use indicates that Clove has several therapeutic properties, such as an aphrodisiac, stomach, carminative, nervous stimulant and tonic. Reports has been documented on their aphrodisiac, antibacterial, antifungal, nematocidal, anti-toothache, anti-oxidant, analgesic, anesthetic and hypotensive properties. Although in Iranian traditional medication several formulation containing clove have been used as a nervous stimulant and cognitive enhancer. (Mohammad, *et al.,* 2009).

2. MATERIALS AND METHODS

2.1. Plant material

The flower bud of *Syzygium aromaticum* (clove) purchased from local market of Chennai and air dried then homogenized to fine powder and stored in airtight bottles.The buds were identified and authenticated by Dr. Rathinakumar, Department of Botany, Presidency College, Chennai.

2.2. Preparation of extracts

Extract from the plant materials were prepared using different solvents such as ethanol, methanol, chloroform and ethyl acetate. Extracts was prepared by using a Soxhlet apparatus as per the standard procedure (Carter, *et al*, 1986).The filtrate was concentrated and the resulting residue was dried, weighed and stored for further use.

2.3. Test organisms

The test organism includes *Staphylococcus aureus, Vibrio choler* and *Pseudomonas aeruginosa, Salmonella typhi, Escherichia coli* and *Proteus mirabilis*

2.4. Determination of the antibacterial activity

The extract were reconstituted in small amount of the corresponding solvent and dissolved in 10% DMSO. The MIC was determined as per the procedure (Irobi *et al*., 1993).The *in vitro* susceptibility testing by tube dilution method against *Staphylococcus aureus, Vibrio cholera, Pseudomonas aeruginosa, Salmonella typhi, Escherichia coli* and *Proteus mirabilis.* Sterile MHB was added to each of the test tube and plant extract was incorporated to the first tube and was serially diluted to obtain a concentration ranging from 32 mg/ml to 0.125 mg/ml. 20µl of the inoculum suspension whose turbidity was matched with McFarland standard were added to each of test tube. The tubes were incubated at 37°C for 24 hours and the tubes were examined for the growth of the organism by observing the turbidity. The MIC was recorded as the Minimum Inhibitory concentration of the extract that suppressed the visible growth of the organism. The tubes without the extract served as control.

2.5. Determination of minimum bactericidal concentration (MBC)

MBC were determined by plating each of the dilution from the tubes on to fresh MHA plates (extract free media). The plates were incubated at 37°C for 24 h and the readings were recorded. MBC was determined as the lowest concentration of the extract that killed the isolate on the solid media.

2.6. Phytochemical analysis

The plant extracts were screened for the presence of steroids, cardiac glycosides, alkaloids, Saponins, tannins, flavonoids, and terpenoids. The presence of phytochemicals in Chloroform and Ethanolic extract were determined by following the procedure of Edeoga *et al.* (2005) and Ajaji *et al.* (2011).

3. RESULTS

Extract from the flower bud of *Syzygium aromaticum* (clove), were prepared using four different solvents such as ethanol, methanol, chloroform and ethyl acetate. Extract were tested for new antimicrobial properties on *Staphylococcus aureus, Vibrio cholera, Pseudomonas aeruginosa, Salmonella typhi, Escherichia coli* and *Proteus mirabilis*.The MIC and MBC value obtained for the extract and the tested bacterial species are shown in table 1. The inhibitory activity of the alcoholic extract was comparatively higher than that of chloroform and ethyl acetate extracts. The extracts inhibited *Escherichia coli and Vibrio cholera* at a MIC value of 4 mg/ml, while showed a MIC value of 16 to 32 mg/ml against *Staphylococcus aureus and Pseudomonas aeruginosa.* None of the extracts inhibited *Salmonella typhi.* .The inhibitory activities of the methanolic and ethanolic extract were comparatively less than that of chloroform and ethyl acetate. *Escherichia coli* and *Vibrio cholera* were inhibited at 2 to 4 mg/ml while *Staphylococcus aureus* and *Pseudomonas aeruginosa* were inhibited at the MIC value from 2 to 32 mg/ml .The MBC values were slightly higher than that of MIC value.

The results of phytochemical analysis of the Chloroform and Ethanol extracts of the clove bud showed the presence of alkaloids, flavonoids, tannins, Saponins and cardiac glycosides (Table 2)

Table 1: Antibacterial activity of different extracts of *Syzygium aromaticum* (clove) flower bud

S.No.	Solvents used	Bacterial species tested and MIC and MBC values in mg/ml											
		Staphylococcus aureus		*Pseudomonas aeruginosa*		*Escherichia coli*		*Vibrio cholera*		*Salmonella typhi*		*Proteus mirabilis*	
		MIC	MBC	MIC	MBC	MIC	MBC	MIC	MBC	MIC	MBC	MIC	MBC
1.	Ethanol	16	32	-	-	4	8	4	8	-	-	8	16
2.	Methanol	2	4	16	32	4	8	2	4	-	-	8	16
3.	Chloroform	4	8	8	16	8	16	4	8	-	-	4	8
4.	Ethyl acetate	16	32	16	32	8	16	4	8	-	-	4	16

Table 2: Phytochemicals present in *Syzygium aromaticum*

Plant Material	Solvent	Phytochemicals analysed						
		ST	F	A	S	T	CG	TE
Flower bud	Chloroform	-	+	-	+	-	+	-
	Ethanol	-	+	+	+	+	-	-

ST: Steroids, F: Flavonoids, A: Alkaloids, S: Saponins, T: Tannins, CG: Cardiac Glycoside and TE: Terpenoids.

4. DISCUSSION

Properties of spices include bactericidal, bacteriostatic, fungistatic, anthelmintic, medicinal and flavouring. There are documented reports are available on the study of antibacterial activity of clove bud oil. In the present study the ethanolic, methanolic, chloroform and ethyl acetate extracts of the clove bud were tested for their antibacterial activity against *Staphylococcus aureus, Salmonella typhi, Escherichia coli, Vibrio cholera, Pseudomonas aeruginosa* and *Proteus mirabilis*. The MIC values ranged from 2 to 32 mg/ml and none of the extracts inhibited *Salmonella typhi*. The alcoholic extracts showed more inhibitory activity than the chloroform and ethyl acetate extracts of the clove bud.

The inhibitory activity of the alcoholic extracts may be due to the extraction of phytochemicals such as flavonoids, cardiac glycosides and Saponins. Sabita Pokhrel *et al.,* (2012) found the activity of ethanolic extracts of clove against *Escherichia coli* was significant. In the present study also, *Escherichia coli* was inhibited at a lesser concentration of the alcoholic extracts of the clove bud. *Staphylococcus aureus* was inhibited at a higher concentration of the extracts and this was in agreement with earlier studies of Sofia *et al*., (2007). The activity may be attributed to the presence of eugenol, a phenolic compound that renders the extracts effective against the species tested. The extracts of the clove buds showed no inhibition on *Salmonella typhi,* while Mostafizur Rahman *et al.,* (2011) has reported the inhibitory activity of the alcoholic extracts of clove on *Salmonella typhi* and the reported MIC value was 420 mg/ml.

5. CONCLUSION

In the present study the tested range of concentration of the extracts was lesser. The alcoholic extracts showed more inhibitory activity and may be attributed to the presence of phytochemicals such as flavonoids, alkaloids, Saponins and tannins present in the extracts.

REFERENCE

Aibinu, I., Adenipekun, T., Adelowotan, T., Ogunsanya, T., Odugbemi, T., 2007. Evaluation of the antimicrobial properties of different parts of *Citrus aurantifolia* (Lime fruit) as used locally. *Afr. J. Trad. CAM*, 4: 185-190.

Ajaji, I., Ajibade, O., Oderinde, R., 2011. Preliminary phytochemical analysis of some plant seeds. *Research Journal of Chemical Science*. 1(3): 58-62.

Amit, R., Shailendra, S., 2006. Ethnomedicinal approach in biological and chemical investigation of phytochemicals as antimicrobials. *http://www. phamainfo.net.*

Carter, T.E., Shanmugasundaram, S., Edamame. 1993. The vegetable soybean. In underutilized crop: pulses and vegetables. *Int J Cur Bio Med Sci*. 6: 219-239.

Edeoga, H.O., Okwu, D.E., Mbaebie, B.O., 2005. Phytochemical ConstituentsOf Some Nigerian Medicinal plants. *African Journal of Biotechnology*. 4(7): 685-688.

Irobi, A. Lade, 1993. Antimicrobial activities off crude leaf extract of Acalypha wilkesiana. *J. Ethnopharmacol.* 39: 171 -174.

Mohammad Hussein Dashti, R., Abbas Morshedi, 2009. The effects of *Syzygium aromaticum* (clove) on learning and memory in mice. *Asian Journal of Traditional Medicines*. 4(4): 128-133.

Mostafizur Rahman, Atikur Rahman, Soriful Islam, Firoz Alam., 2011. *In vitro* controlling of selected Human diarrhoea causing bacteria by clove extracts. *International Journal of Biomolecules And Biomedicines*, 1(2): 17-26.

Papachan Karur Sofia, Rajendra Prasad, Virendra kumar Vijay, Ashok Kumar Srivastava, 2007. Evaluation of antibacterial activity of Indian spices against common Foodborne pathogens. *International Journal of Food Science and technology*, 42: 910-915.

Sabita Pokhrel, Ratan singh, Preeti Gautam, Vijay Kant Dixit, Amar Jyoti Das, 2012. Comparison of antimicrobial activity of crude ethanolic extracts and essential oils of spices against five strains of diarrhoea causing *Esheriachia coli. International Journal of Pharmacy and Life science,* ISSN-0976-7126.

15

Preliminary Phytochemical Analysis of Red Purple Fruit of *Cocculus hirsutus* Linn.

Rakkimuthu, R. and Aravinthan, K.M.

1. INTRODUCTION

Phytochemicals are compounds present in fruits that are not required for normal functioning of the body, but have a beneficial effect on health or play an active role in amelioration of diseases. Phytonutrients have various health benefits like antimicrobial, anti-inflammatory, cancer preventive, antidiabetic and antihypertensive. The pigments of the fruits have many therapeutic benefits including vasoprotective and anti-inflammatory properties, antiproliferative, chemo protective and anti-neoplastic properties, reversing age related deficits (Joseph *et al.*,1999) and useful in controlling oxidative stress during pregnancies complicated by intrauterine growth retardation (Pawlowicz *et al.*,2000). *Cocculus hirsutus* Linn. is a climber and widely distributed in tropical and subtropical climates. This plant is commonly called as kattu kodi in Tamil. It belongs to the family Menispermaceae. The fruit is a drupe which is size of small pea with red purple endocarp (Fig.1a and Fig.1b). The plant is used in treatment of gonorrhea, spermatonorrhea, Urinary troubles, diarrhea and hyperglycemia (Kirtikar and Basu, 2002). The leaves of the plant have been evaluated for anti hyperglycemic (Badole *et al.*, 2006); antibacterial (Panda *et al.*, 2007); diuretic and laxative (Ganapaty *et al.*, 2002). Several isoquinoline alkaloids are present in leaves and roots; the important ones cohirsine, cohirsinine, cohirsitinine, hirsutine and jamtine (Viqaruddin *et al.*, 1987; 1991; 1992; 1993). Since no detail scientific approach has been made to study phytochemicals of *cocculus hirsutus* fruit. So in the present investigation preliminary phytochemical constituents of methanol extract of *cocculus hirsutus* fruit was screened.

Fig. 1a

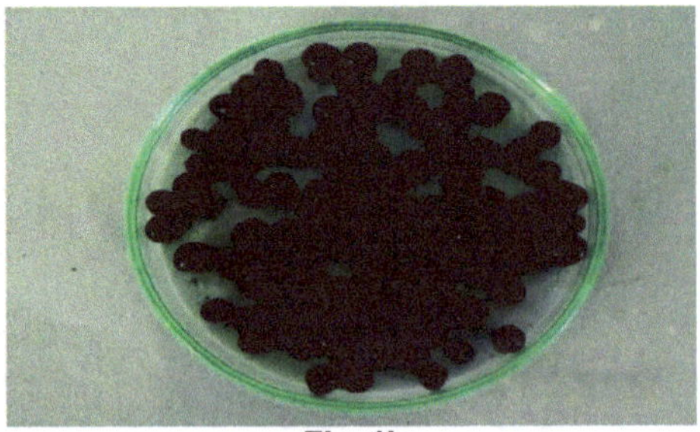

Fig. 1b

2. MATERIALS AND METHODS

2.1. Sample collection

Fresh *Cocculus hirsutus* fruits were sampled places from in and around of coimbatore , Tamilnadu, India during the month of November 2011. After sampling, the fruits were immediately transported to laboratory, put in plastic bags, and stored at -20°C until use. The plant was identified at the Department of Botany, NGM College, Pollachi, Coimbatore, Tamilnadu, India, as *cocculus hirsutus* Linn.

2.2. Preparation of Extracts

The fruit of *cocculus hirsutus* L were first washed carefully under running tap water, dried and meshed with a tissue paper. Then, the homogenized fruit sample of 10 gm was weighted in a beaker and 200 ml of methanol. The samples were stirred with a glass rod for every 5 minutes to ensure a well extraction. After the extraction in the dark at room temperature for 4 h, the extract was centrifuged (5000 rpm/min) at 4°C for 20 minutes in a Remi (India). Then the supernatant was separated and filtered. The filtered extract was evaporated under reduced pressure and the evaporated sample was used for further analysis.

2.3. Qualitative phytochemical analysis

Chemical tests were carried out on the methanolic fruit extract for the qualitative determination of phytochemical constituents as described by Harborne (1973), Trease and Evans (1989) and Sofowara (1993).

2.3.1. Alkaloids

2 ml of fruit extract was divided into two portions. Mayer's reagent was added to one portion and Draggendoff's reagent to the other. The formation of a cream (with Mayer's reagent) or reddish brown precipitate (with Draggendoff's reagent) was regarded as positive for the presence of alkaloids.

2.3.2. Flavonoids

5 ml of dilute ammonia solution were added to a portion of the methanol filtrate of fruit extract followed by addition of concentrated H_2SO_4. A yellow colouration observed in the extract indicated the presence of flavonoids.

2.3.3. Saponins (Frothing test)

To 0.5 g of extract was added 5 ml of distilled water in a test tube. The solution was shaken vigorously and observed for a stable persistent froth. The frothing was mixed with 3 drops of olive oil and shaken vigorously. An appearance of creamy mass of small bubbles indicated the presence of saponins

2.3.4. Steroids

Two milliliter of acetic anhydride was added to 0.5 g of methanol extract of each sample with 2 ml H_2SO_4. The colour changed from violet to blue or green, indicating the presence of steroids.

2.3.4. Tannins

About 0.5 g of the extract was boiled in 10 ml of water in a test tube and then filtered. A few drops of 0.1% ferric chloride was added and observed for brownish green or a blue-black colouration indicating the presence of tannins.

2.3.5. Terpenoids (Salkowski method)

To 0.5 g each of the extract was added 2 ml of chloroform. Concentrated H_2SO_4 (3 ml) was carefully added to form a layer. A reddish brown colouration of the interface indicated the presence of terpenoids.

2.3.6. Cardiac glycosides (Keller-Killiani test)

To 0.5g of extract diluted to 5 ml in water was added 2 ml of glacial acetic acid containing one drop of ferric chloride solution. This was underplayed with 1 ml of concentrated sulphuric acid. A brown ring at the interface indicated the presence of deoxysugar characteristic of cardenolides. A violet ring may appear below the brown ring, while in the acetic acid layer a greenish ring may form just above the brown ring and gradually spread throughout this layer.

3. RESULTS AND DISCUSSION

The phytochemical investigation of methanol extract of *Cocculus hirsutus* leaves showed the presence of carbohydrates, glycosides, flavonoids, tannins, amino acids and saponins (Panda *et al*, .2009 and Satish *et al.*, 2009). In our present study, the qualitative phytochemical analysis was done by using colour forming and precipitating chemical reagents to generate preliminary data on the constituents of the plant extract.

Table 1: Result of preliminary phytochemical analysis

S.No.	Phytoconstituents	Observation	Methanol extract
1.	Alkaloid	No colour change	-
2.	Flavonoids	Yellow colour formed	+
3.	Saponin	No froth form	-
4.	Steroids	No colour change	-
5.	Tannins	Brown-green coloured formed	+
6.	Terpenoids	No colour change	-
7.	Cardiac glycosides	No colour ring form	-
	+ Present; - Absent		

The chemical tests revealed the presence of tannins and flavonoids in the methanolic fruit extracts of *Cocculus hirsutus* L. The results of preliminary qualitative phytochemical analysis are tabulated.

4. CONCLUSION

Based on the result, it was concluded that methanol extract of *Cocculus hirsutus* L fruit contain tannins and flavonoids. Further studies are required to identify the pigmented compounds and its antioxidant and anticancer activity of this fruit.

REFERENCES

Badole, S., Patel, N., Badhankar, S., Jain, B. and Bharadwaj, S. 2006. Antihyperglycemic Activity of aqueous extract of leaves of *Cocculus hirsutus* (L) Diels in alloxan induced diabetic mice. *Ind. J. of Pharmacol.*, 38(1): 49-53.

Ganapaty, S., Dash, G.K., Subburaju, T. and Suresh, P., 2002. Diuretic, laxative and toxicity study of *Cocculus h)irsutus* aerial parts. *Fitoterapia,* 73(1): 28-31.

Harborne, J.B., 1973. Phytochemical Methods: A Guide to Modern Technique of Plant Analysis. Chapman and Hall Ltd. London: 49-188.

Joseph, J.A., Shukitt-Hale, B., Denisova, N.A., Bielinski, D., Martin, A., McEwen, J.J. and Bickford, P.C., 1999. Reversals of age-related declines in neuronal signal transduction cognitive and motor behavioral deficits with blueberry, spinach or strawberry dietry supplementation. *The Journal of Neurosciences*. 8114-8121.

Kirtikar, K.R. and Basu, B.D., 2002. *Indian Medicinal Plants*, Sri Satguru Publication, New Delhi, 3rd revised enlarged edition. 1: 120.

Panda, B.R., Mohanta, S.R., Misra, U.S., Kar, S., Panda, B.K. and Chakraborty, P., 2007. Antimicrobial activity of the leaves of *Cocculus hirsutus*. *Indian Drugs*. 44(2): 108-110.

Panda Bikash Ranjan, Mohanta, S.R., Sii, S. and Das, G.K., 2009. Preliminary phytochemical analysis and wound healing activity from the leaves of *Cocculus hirsutus*. *International Journal of Pharmaceutical Sciences and Nanotechnology*. 2(1): 675-678.

Pawlowicz, P. and Stachowiak, G., 2000. Administration of natural anthocyanins derived from chokeberry retardation of idiopathic and preclamptic origin. Influence on metabolism of plasma oxidized lipoproteins: the role of autoantibodies to oxidize low density lipoproteins. *Ginekol Pol.* 71: 848-853.

Satish, V., Ravichandrian, V.D., Usha Gavani, Padmaa M. Paarakh, 2009. Antimicrobical studies on the extract of *Cocculus hirsutus* and *Hyptis suaveolens*. *Indian Journal of Natural Products and Resources*. 1: 49-52.

Sofowora, L.A., 1993. Medicinal plants and traditional medicine in Africa. Spectrum Books Ltd, Ibadan, 55-71.

Trease, G.E. and Evans, W.C., Pharmacognosy: 13th ed, Bailliere Tindall, London.

16

Biological Potency of Medicinal Plant Extracts on Life Stages of *Spodoptera litura* (Fab.) (Noctuidae: Lepidoptera)

Mariappan Suganthy, Sentrayaperumal Sundareswaran, Periyaswamy Sakthivel and Lakshmanan Nalina

1. INTRODUCTION

Mounting pressure of teeming population and shrinking land resources have necessitated the production of increasingly more food per unit area. Insect pests, diseases and weeds are important biotic constraints inflicting 20-25% loss in agriculture production. Synthetic pesticides have of course, played a very significant role in restricting many pest problems. However, their indiscriminate has been its toll by creating several problems like pesticide resistance in insects, contamination of food with toxic residues, resurgence of pests and deleterious effects on non target organisms. The number of effective pesticides available is dwindling day by day further having less choice among which to choose. These ill effects have arised interest in alternate methods of plant protection.

The extent to which various plant species are utilized for pest control depends, not only on their effectiveness against target pests, but also on socio-economic, environmental and policy considerations. Keep these in mind the present study on the efficacy of botanicals against *Spodoptera litura* was carried out.

2. MATERIALS AND METHODS

2.1. Mass Culturing of *Spodoptera litura*

The culture of *Spodoptera litura* was maintained in the artificial semi-synthetic diet and also in castor leaves in the laboratory at 25 ± 1° C and 75 ± 5 % RH.

2.2. Extraction of Plant Material

The leaves of *Adathoda vasica, Vitex negundo, Catharanthus roseus, Trachyspermum ammi, Chrysanthemum coronarium, Curcuma longa, Acorus calamus* were collected. Ten per cent aqueous extracts of these plant parts were prepared taking 100 grams of these leaf material, making them into paste and volume made into 1 litre using distilled water. The extracts were kept as such over night and filtered after 24 hours using filter paper. The extracts prepared were labeled and stored in the refrigerator. Apart from plant extracts, three TNAU neem formulations *viz*., TNAU NO (Neem Oil) 60 EC (A), TNAU NO (Neem Oil) 60 EC (C) and TNAU NOPO (Neem Oil + Pungam Oil) 60 EC (5%) were also used. Treatments one to seven were 10 per cent plant extracts while T_8, T_9 and T_{10} were 5 per cent solutions.

2.3. Bio-assay

For the bio-efficacy test against *S. litura* the aqueous extracts were sprayed on the potted castor seedlings of 45 days old and considered as a treatment. To each plant, ten *S. litura* early 3rd instar larvae were released and covered using mylar film cage. There were ten treatments and each treatment was replicated thrice. The larvae were observed for mortality at 6, 12, 24, 48, 60 and 72 hours after release (HAR). To assay the antifeedant activity, leaf discs of 9 cm diameter were punched from castor leaves and dipped in aqueous solutions of plant extract for 5 seconds. The leaf discs were shade dried and placed in a glass petridish, which had a filter paper over moist cotton wool. Leaf discs dipped in water were maintained as check. Ten early third instar larvae were starved for 2 hours, placed over the leaf disc and allowed to feed at room temperature. Mortality count was taken at 12, 24, 48, 60 and 72 hours after treatment. After 72 hours, the larvae were transferred to fresh diet and observed till pupation and percentage pupation was worked out. Larval-pupal malformation was observed. After pupation, the pupae were observed till moth emergence for pupal-adult malformation. Percentage larval-pupal or pupal-adult malformations were worked out. The pupae of individual treatments were collected and kept in petriplate over saw dust in moth-emergence cage. The leaves of *Nerium odorum* were placed in cages with the petioles immersed in water for oviposition. These leaves were uniformly sprayed with plant extracts and air dried. This set up was placed inside the moth-emergence cage into which the pupae of that particular treatment were placed already. The emerged adults were provided with 10 per cent sucrose solutions. Then the moths started laying eggs on the nerium leaf. At the end of egg laying period, number of egg mass laid in each treated set was counted and recorded. The eggs laid over the treated leaf surface were observed daily till the eggs hatched out to assess the hatchability of eggs in each treatment. Percentage hatchability of eggs was calculated.

2.4. Statistical Analysis

The data collected in the experiments were statistically analyzed using completely randomized design. Percentage data in the experiment were converted into corresponding angles (arcsine percentage) for statistical interpretation. Whenever there was zero value, data were transformed into square root for statistical interpretation. Duncan's Multiple Range Test (DMRT) was applied to analyze the data. In the table, in each column, means followed by a common letter were not significantly different at 5 per cent level.

3. RESULTS AND DISCUSSION

Results of bio-efficacy studies revealed that at 12 and 24 HAT, the treatments were found to have no effect on the larvae. At 48 HAT, all the treatments were found to be non-significant. At 60 HAT, only TNAU neem formulations were found to be effective with 13.33 per cent larval mortality and all other treatments were found to be on par with control. At 72 HAT, NO 60 EC [A] was found to be superior (40.00), which was on par with NO 60 EC [C], NOPO 60 EC, *Acorus calamus* and *Vitex negundo*, followed by 13.33 per cent in *Adathoda vasica*, *Catharanthus roseus, Chrysanthemum coronarium* and *Curcuma longa*. Minimum mortality of 0.007 per cent was observed in *Trachyspermum ammi* which is found to be on par with control. This is mainly because of antifeedant property. Percentage pupation was observed to be minimum in TNAU NO 60 EC [A] (40.00) followed by 53.33 per cent in NO 60 EC [C] and NO PO 60 EC. Maximum percentage of pupation was observed with *Trachyspermum ammi* (99.997) which was found to be on par with control. Though *Acorus calamus* and TNAU NO 60 EC [C] and NOPO 60 EC recorded 13.33 per cent malformation, all the treatments were found to be non significant. Minimum number of egg masses were observed in *Curcuma longa* treatment (2.00), followed by TNAU neem formulation, due to oviposition deterrent nature of neem products. Regarding hatchability, all the treatments were found to be non significant. The treatments have no effect on the hatchability of eggs (Table 1).

Hermawan *et al.* (1994) reported antifeedant and antioviposition activities of the fractions of extract from a tropical plant. *Andrographis paniculata*, against the diamondback moth, *Plutella xylostella.* Sahayaraj and Sekar (1996) and Sahayaraj (1998) recorded antifeedant effect of some plant extracts on *Spodoptera litura*. Hiremath *et al.* (1997) reported the insecticidal activity of some Indian plant extracts against *Nilaparvatha lugens*. Kulkarni *et al.* (1997) reported that some medicinal

Table 1: Bio-efficacy of medicinal plant extracts on *Spodoptera litura* life cycle

	Treatments	Larval mortality [%]			Pupation [%]	Pupal or adult malformation [%]	No. of egg mass/treatment	Hatchability [%]
		48 HAT	60 HAT	72 HAT				
T_1	*Adathoda vasica*	0.03^{b}(0.19)	0.03^{b}(0.19)	13.33^{b}(17.71)	86.67^{cd}(72.28)	6.66^{a}(8.85)	4.00^{bc}(11.47)	99.97^{a}(89.80)
T_2	*Vitex negundo*	0.03^{b}(0.19)	0.03^{b}(0.19)	20.00^{ab}(26.56)	80.00^{bc}(63.47)	6.66^{a}(8.85)	2.66^{ab}(9.26)	99.97^{a}(89.80)
T_3	*Catharanthus roseus*	0.03^{b}(0.19)	0.03^{b}(0.19)	13.33^{b}(17.71)	86.67^{cd}(72.28)	0.03^{a}(0.19)	5.66^{bc}(13.68)	99.97^{a}(89.80)
T_4	*Trachyspermum ammi*	0.03^{b}(0.19)	0.03^{b}(0.19)	0.07^{c}(0.27)	99.97^{d}(89.80)	0.03^{a}(0.19)	4.33^{bc}(11.99)	99.97^{a}(89.80)
T_5	*Chrysanthemum coronarium*	0.03^{b}(0.19)	0.03^{b}(0.19)	13.33^{b}(17.71)	86.67^{cd}(72.28)	0.03^{a}(0.19)	4.33^{bc}(11.89)	99.97^{a}(89.80)
T_6	*Curcuma longa*	0.03^{b}(0.19)	0.03^{b}(0.19)	13.33^{b}(17.71)	86.67^{cd}(72.28)	0.03^{a}(0.19)	2.00^{a}(6.55)	99.97^{a}(89.80)
T_7	*Acorus calamus*	0.03^{b}(0.19)	0.03^{b}(0.19)	26.66^{a}(30.78)	73.33^{abc}(59.21)	13.33^{a}(17.71)	5.00^{bc}(12.87)	99.97^{a}(89.80)
T_8	TNAU- NO 60 EC [A]	6.66^{a}(8.85)	13.33^{a}(17.71)	40.00^{a}(38.85)	40.00^{a}(38.85)	6.66^{a}(8.85)	3.33^{abc}(10.76)	99.97^{a}(89.80)
T_9	TNAU- NO 60 EC [C]	0.03^{b}(0.19)	13.33^{a}(17.71)	33.33^{a}(34.63)	53.33^{ab}(46.93)	13.33^{a}(17.71)	3.67^{abc}(10.76)	99.97^{a}(89.80)
T_{10}	TNAU- NOPO 60 EC	0.03^{b}(0.19)	13.33^{a}(17.71)	33.33^{a}(29.99)	53.33^{ab}(47.29)	13.33^{a}(17.71)	3.00^{abc}(9.88)	99.97^{a}(89.80)
T_{11}	Untreated control	0.03^{b}(0.19)	0.03^{b}(0.19)	0.03^{c}(0.19)	99.97^{d}(89.80)	0.03^{a}(0.19)	6.33^{c}(14.56)	99.97^{a}(89.80)

* The values are mean of three replications; * Figures in parentheses (except No. of egg mass) are arc sin transformed.
* Figures in parentheses (No. of egg mass) are square root transformed.

plant extracts have insecticidal action on *Albizia* defoliator, *Rhesala imparata*. Antifieedant and insecticidal activity of some medicinal plant extracts against *Dalbergia sissoo* defoliator *Plecoptera reflexa* was reported by Meshram (2000).

REFERENCES

Hermawan, W., Kajiyama, S., Tsukuda, R., Fujisaki, K., Kobayashi, A., Nakasuji, F., 1994. Antifeedant and antioviposition activities of the fractions of extract from a tropical plant, *Andrographis paniculata* (Acanthaceae), against the diamond back moth, *Plutella xylostella* (Lepidoptera: Yponomeutidae). *Applied Entomology and Zoology*, 29(4): 159-166.

Kulkarni, N., Joshi, K.C., 1997. Insecticidal action of some plant extracts against *Albizia* defoliator, *Rhesala imparata* Walker (Lepidoptera: Noctuidae). *Entomon*, 22(2):135-139.

Meshram, P.B., 2000. Antifeedant and insecticidal activity of some medicinal plant extracts against *Dalbergia sissoo* defoliator *Plecoptera reflexa* Gue. (Lepidoptera : Noctuidae). *Indian Forester*, 126(9):961-965.

Sahayaraj, K., 1998. Antifeedant effect of some plant extracts on the Asian armyworm, *Spodoptera litura* (Fabricius). *Current Science*, 74(6): 523-525.

Sahayaraj, K., Sekar, R., 1996. Efficacy of plant extracts against tobacco caterpillar larvae in groundnut. *International Arachis Newsletter*, 16: 38.

17

Evaluation of Antifungal Activity of *Kedrostis rostrata* Cogn. Against *Keratinophilic* Fungi

Logaprabha, V. and Tamil Selvi, K.S.

1. INTRODUCTION

Keratinophilic fungi are the natural colonizers of keratinic substances like hair, nail, fur, feathers, hooves, claws, beaks, scales etc. Cattle sheds, garbage, animal burrows, sewage, birds nest, feather and hair dumping sites, poultry farms, public places like parks, schools, market place etc., are the common habitats of these fungi (Ali-Shtayeh and Rana, 2000). They utilize keratin as the sole source of carbon and nitrogen and produce an extracellular enzyme keratinase.

Keratinophilic fungi have been receiving considerable attention in recent days as these include dermatophytes which are potential pathogens causing various diseases in skin and scalp (Weitzman and Summerbell, 1995). Human exposure to these fungi is a matter of health risk. Several non-pathogenic fungi are now being reported as opportunistic pathogen and their occurrence in various environments, where they are naturally occurring is not directly investigated. Although several antifungal drugs are available in the market, its use is being limited due to a number of reasons such as low potency, poor solubility, development of resistant strains and drug toxicity. Hence there is an essential need to evaluate plant extracts scientifically for their medicinal properties to discover more safer and effective antifungal agents.

The plant *K. rostrata* Cogn. a member of Cucurbitaceae, already been characterized to be effective in the treatment of asthma, urinary tract infection (Giday, 2001), diarrhoea, small pox, skin diseases (Tabuti *et al.*, 2003) and snake bite is now been explored for its antifungal activity. The present study was undertaken to evaluate the antifungal property of different solvent extracts of leaves of the plant *K. rostrata,* exploring its inhibitory nature against the keratinophilic fungi *A. niger*.

2. MATERIALS AND METHODS

2.1. Plant Material

The plant material was collected from local areas of Coimbatore and the reference sample was identified with the Flora of the Presidency of Madras (Gamble, 1915). The plant *K. rostrata* is a prostrate or scandent herb with orbicular or reniform leaves, small flowers and an ovoid, beaked, red fruit.

2.2. Preparation of Extract

The collected plant material was washed with running water to remove all the adhering materials. Then weighed amount of leaf were extracted separately with various extracts like petroleum ether, chloroform, acetone, ethanol and water. The extracts were filtered after 24 hrs and allowed to evaporate under reduced pressure at room temperature. The residues were dissolved in known quantities of distilled water, for screening of anti-fungal activity. Residues were preserved in sealed vials and stored at 4^0C for further use.

2.3. Screening of Anti-fungal activity: (Mukherjee *et al.*, 1995)

The keratinophilic fungus (*A. niger*) was isolated from various soil samples of Coimbatore district. Antifungal activity was tested by the agar well diffusion method. Six wells of 6mm diameter were made using a sterile cork borer. The wells were filled with solvent extracts of different drug concentrations (4µg/well, 8µg/well, 12µg/well and 16µg/well). The standard fungicide Fluconazole (10µg/well) served as a positive control while distilled water was the negative control. The antifungal assay plates were incubated at room temperature for 48 hours. The diameter of the zones of inhibition around each of the wells was taken as a measure of the anti-fungal activity. The experiments were carried out in triplicates and the mean diameter of the inhibition zone was recorded.

2.4. Determination of Minimum Inhibitory Concentration (MIC) and Minimum Fungicidal Concentration (MFC): (Jones *et al.*, 1985)

The extracts which showed antifungal activity in the agar well assay, were subjected to the MIC assay, by micro broth dilution method according to CLSI (Clinical and Laboratory Standard Institute) standard. MIC values were interrupted as the highest dilution of the sample, which showed clear fluid with no development of turbidity. The highest dilution that yielded no growth on the solid medium was taken as MFC.

3. RESULTS

The antifungal activity of different solvent extracts (petroleum ether, chloroform, acetone, ethanol and water) of leaves of *K. rostrata* were determined on the basis of zone of inhibition, produced by the target species *A. niger*. The chloroform and aqueous extracts showed good inhibition while the other extracts did not show any significant inhibition in the test organism. Therefore the antifungal activity of different concentrations of the extracts was determined. It was observed that 16µg concentration of chloroform extract and aqueous extract showed more activity, whereas 4µg showed little activity and no activity was seen in the wells with negative control. The extracts showing maximum activity were further screened for Minimum Inhibitory Concentration (MIC) and Minimum Fungicidal Concentration (MFC) against *A. niger*.

Table 1: Antifungal activity of different solvent extracts of *K. rostrata* on *A. niger*

	C	PE	A	E	W	F	DW
Zone of Inhibition*	+	-	-	-	+	+	-
Inhibition ZoneDiameter (mm)*							
4µg	9	-	-	-	5	21	-
8µg	14				8		
12µg	18				11		
16µg	20				15		
MIC*	4	-	-	-	4	-	-
MFC*	4	-	-	-	4	-	-

*- average of triplicates, C-Chloroform; PE-Petroleum Ether; A-Acetone; E-Ethanol; W-Water; F-Flucanazole; DW-Distilled Water, MIC-Minimum Inhibitory Concentration; MFC-Minimum Fungicidal Concentration.

4. DISCUSSION

In the present study, an attempt was made to analyze the antifungal activity of the various solvent (chloroform, petroleum ether, acetone, ethanol and water) extracts of *K. rostrata*. Amongst the extracts, chloroform and aqueous extracts of *K. rostrata* proved to possess strong antifungal activity against *A. niger*, supporting the findings of Priyavardhini *et al.*, (2012). The other extracts examined did not show any significant inhibition in the test organism. MIC and MFC also determined that in addition to the inhibitory nature, the extracts are also found to possess fungicidal activity. Further, the chloroform extract is found to be equally effective on *A. niger*, compared to the widely used commercial fungicidal antibiotic Fluconazole.

REFERENCES

Ali-Shtayeh, M.S., Jamous Rana, M.F., 2000. Keratinophilic fungi and related dermatophytes in polluted soil and water habitats. *In*: Kushwaha, RKS, Guarro, J. eds. Biology of dermatophytes and other keratinophilic fungi. Revista Iberoamericana de Micologia, Bilbao. 51-59.

Gamble, J.S., 1986. Flora of the Presidency of Madras, volume-1, Botanical Survey of India, Calcutta, India. 540-41.

Giday, M., 2001. An Ethnobotanical study of medicinal plants used by the Zay people in Ethiopia. *CBM:s Skriftserie*, 3: 81-99.

Jones, R.N., Barry, A.L., Gavan, T.L., Washington, J.A., 1985. Susceptibility tests: micro dilution and macro dilution broth procedures. *In*: Manual of Clinical Microbiology, 4th edn. (Eds). E.H. Lennette, A.Balows, W.J. Hausler and J.H.Shadomy. American Society for Microbiology, Washington, pp. 972-977.

Mukherjee, P.K., Giri, S.N., Saha, K., Pal, M., Saha, B.P., 1995. Antifungal screening of *Nelumbo nucifera* (Nymphaeaceae) rhizome extract. *Indian J. Microbiol.* 35: 320-327.

Priyavardhini, S., Vasantha, K., Tresina Soris, P., Mohan, V.R., 2012. Antifungal activity of *Kedrostis foetidissima* (Jacq.) Cogn. International Journal of PharmTech Research, 4 (1): 44-48.

Tabuti, J.R.S., Lye, K.A., Dhillion, S.S., 2003. Traditional herbal drugs of Bulamogi, Uganda: Plants, use and administration. *J. Ethnopharmacol.* 88:19-44.

Weitzman, I., Summerbell, R.C., 1995. The Dermatophytes. *Clin. Microbiol. Rev.* 8(2): 240-259.

18

Wound Healing Chasmophytes of Velliangiri Hills, Southern Western Ghats of Coimbatore District, Tamil Nadu, India

Binu Thomas, Rajendran, A. and Aravindhan, V.

1. INTRODUCTION

India has a rich tradition of plant-based knowledge on healthcare. It is reported that traditional healers use 2500 plant species and 100 species of plants serve as regular sources of medicine (Pei, 2001). Traditional medical knowledge of medicinal plants and their use by indigenous cultures are not only useful for conservation of cultural traditions and biodiversity but also for community healthcare and drug development in the present and future. Medicinal plants have curative properties due to the presence of various complex chemical substances of different composition which are formed as secondary plant metabolites in various parts of the plants such as roots, shrubs, leaves, flowers etc. (Gurib-Fakim, 2006).

A large number of plants are equally used by tribals and folklore traditions in India for treatment of cuts, wounds and burns (Biswas and Mukherjee, 2003). A wound is a type of injury in which skin is torn, cut or punctured, that disturbance in the normal skin anatomy and function. Wound healing consists of orderly progression of series of events that establish the integrity of damaged tissue and thereby the affected parts repair itself after injury. The wound healing herbs encourage blood clotting and induce the healing of wounds (Sudhakar *et al.*, 2003; Nagori and Solanki, 2011; Binu Thomas *et al.*, 2012).

The rock crevices in which more finely grained components and accumulate than on the surface to produce a somewhat copious vegetation, that of chasmophytes (Binu Thomas *et al.*, 2012). While studying chasmophytic diversity of the Velliangiri hills, a preliminary ethno-botanical survey on tribe *Malasars* demonstrated that wounds were one of the major problems among these tribals and they prepared herbal medicines with a number of plants to heal wounds. The present study was performed with the aim of producing an inventory of the plants used by traditional healers in Velliangiri Hills to document the traditional therapies practiced for various wound and related injuries.

2. MATERIALS AND METHODS

2.1. Study area and tribe *Malasars*

A Velliangiri hill is floristically very rich and socio-religiously important since a famous temple called '*Velliangiri Aandava*r' temple also called "*Thenkailayam*" (South Kailas) is situated at the peak of hills (1840 m above MSL) and it form a major hill range in Nilgiri Biosphere Reserve (NBR). The geographical position of the study area is lies between the longitude 6°- 40` and 7°- 10` E and latitude 10°-55` and 11°-10 N with altitudes varies from 520 and 1840 m above MSL. The average rain fall in the hill is ca 3500 mm at the foot hills and 4500 mm at the peak per year respectively. It forms a part of Western boundary of Coimbatore District, Tamil Nadu, bordering the Palakkad District of the Kerala (Aravindhan *et al.*, 2011; Binu Thomas *et al.*, 2012) (Plate-1).

Malasars are one of the limited and primitive forest tribes, their settlements are situated near Poondi. They are illiterates and speak tamil and they reside in hamlets know as pathes formed of huts made of plasted bamboo with mud. They have a very good knowledge about the medicinal plants and their uses. Several field visits were conducted during September 2011 – October 2012 to collect information on chasmophytic medicinal plants used by *Malasars* tribes for curing cuts and wounds in their daily life activities. During the survey queries were made on the various species of plants used regularly and occasionally as to cure cuts and wounds (Plate-1).

After collecting required data, identified their botanical identity with help of persisted *Flora of the Presidency of Madras* (Gamble and Fischer, 1915-1931), *The Flora of Tamilnadu Carnatic* (Matthew, 1983) and also authentificated with the specimens deposited in the Herbarium of Bharathiar University (BUH), Coimbatore, Tamil Nadu.

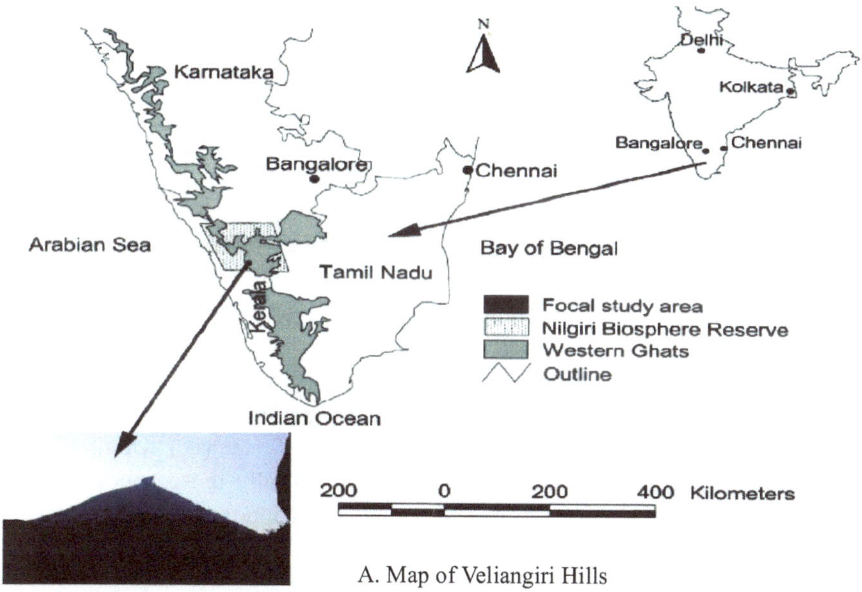

A. Map of Veliangiri Hills

B. Malasars settlements in study area

C. Collecting traditional knowledge from old tribal informant

3. RESULT & DISCUSSION

The present study includes information on 45 plant species belonging to and 28 families and 41 genera were used by *Malasars* tribe of Velliangiri hills of Coimbatore district of Tamil Nadu for the treatment of cuts and wounds (Table- 1). Among the 28 families of medicinal plants represented, Asteraceae is first dominant family with 7 species, followed by Acanthaceae and Rubiaceae with 4 species, Commelinaceae with 3 species, Fabaceae, Pedaliaceae and pteridaceae with 2 species each. The total of 45 plants collected, 7 species from Pteridophytes.

Malasars are using various parts of the plant like leaves, roots and rhizome, stem, aerial parts, seeds etc. for medicinal preparations. Among them, mostly the whole plant extract or paste is used (19 species) followed by leaves (18 species), roots (2 species), aerial parts, stem, succulent parts, rhizome, bulbs and seeds having 1 species each. (Fig.1). The wound healing property of *Tridax procumbens* L. was confirmed through pharmacological studies by Mundada and Shivhare (2010). The extracts of the different parts of the pteridophytic chasmophytes such as *Actiniopteris radiata* (Sw.) Link., *Cheilanthes tenuifolia* (Burm.f) Sw., *Pteris confusa* T.G., *Pteris vittata* L. and *Pteridium aquilinum* (L.) Kuhn. Were is applied externally for curing cuts and wounds. Usually these tribes are using whole plant juice or leaf extract over the cuts and wounds.

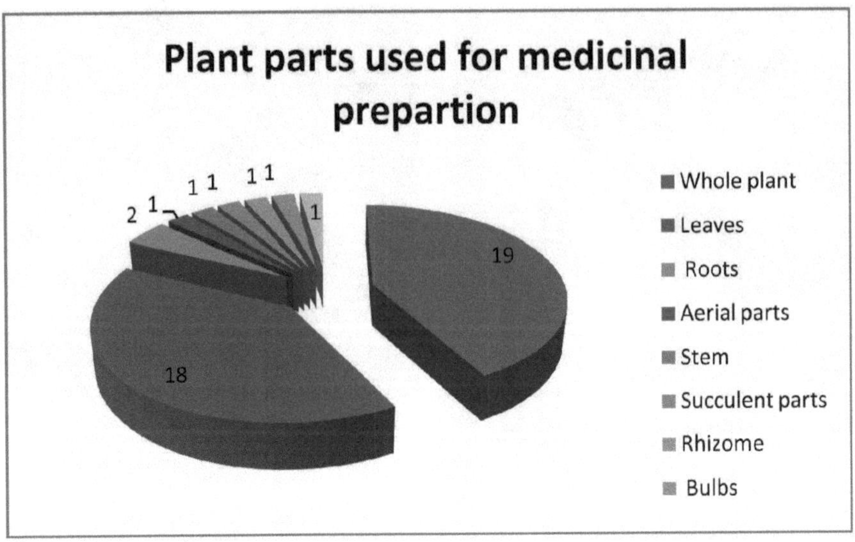

Fig. 1: The various plant parts used for medicinal preparation by *Malasars*

Table 1: List of chasmophytic medicinal plants which are used for wound healing purposes by tribe *Malasars* of Vellinagiri Hills.

Sl No.	Botanical Name	Family	Parts used	Mode of drug preparation
1.	*Actiniopteris radiata* (Sw.) Link	Actiniopteridaceae	Whole plant	Plant paste is applied over the cuts and wounds.
2.	*Ageratum houstonianum* Mill.	Asteraceae	Leaves	Leaves are crushed along with lime and applied on cuts caused by iron knives;
3.	*Alysicarpus monilifer* (L.) DC.	Fabaceae	Leaves	Leaf paste is applied on freshly wounded cuts
4.	*Aristida setacea* Retz.	Poaceae	Whole plant	Whole plant parts are mixed with the stem bark of *Pongamia pinnata* and the mixture is made into a paste and applied topically on affected places to heal wounds.
5.	*Barleria buxifolia* L.	Acanthaceae	Root	Root paste is applied on boils and burns and swellings
6.	*Barleria cuspidata* Heyne *ex* Nees	Acanthaceae	Root	Root paste is applied on boils and swellings
7.	*Bidens pilosa* L.	Asteraceae	Whole Plant	The warm juice of the fresh plant is used for cuts and wounds
8.	*Blechnum orientale* L.	Blechnaceae	Leaves	Consumption of juice extracted from the leaf is used to cure intestinal wounds
9.	*Boerhavia diffusa* L.	Nyctaginaceae	Leaves	Leaf paste is applied on the cuts and wounds to stop bleeding.
10.	*Canscora diffusa* (Vahl.) R. Br.	Gentianaceae	Leaves	Leaves mixed with a pinch of calcium, ground in to paste and applied on cuts for one time a day for 4 days
11.	*Chamaecrista absus* (L.) Irwin & Barneby	Caesalpiniaceae	Leaves	Crushed leaf paste is applied to wounds.
12.	*Cheilanthes tenuifolia* (Burm.f) Sw.	Cheilanthaceae	Leaves	Fronds cut into pieces, made to a paste and it is applied on abscess
13.	*Commelina benghalensis* L.	Commelinaceae	Whole Plant	Juice extracted from the stem is mixed with the stem juice of *Canna indica* and fruits of *Areca catechu*. The mixture is applied topically on affected places to heal wounds.
14.	*Commelina ensifolia* R. Br.	Commelinaceae	Aerial parts	Water extract of fresh aerial part is applied externally to heal burn injuries, itches and boils.
15.	*Cyanotis villosa* (Spreng.) Schult.f	Commelinaceae	Stem	Stem of this plant and stem juice of *Canna indica* are ground into a paste and applied externally to heal wounds.
16.	*Desmodium triflorum* (L.) DC.	Fabaceae	Whole Plant	Fresh leaf paste is applied to wounds and abscess.
17.	*Didymocarpus humboldtiana* Gard.	Gesneriaceae	Whole Plant	Whole parts of this plant, stem bark of *Pongamia pinnata* and leaf of *Abutilon*

(Contd.)

Sl No.	Botanical Name	Family	Parts used	Mode of drug preparation
				indicum are ground into a paste and applied topically on affected places to heal wounds.
18.	*Ecbolium viride* (Forssk.) Alston.	Acanthaceae	Leaves	Leaf juice is poured to cure ear boils.
19.	*Elephantopus scaber* L.	Asteraceae	Leaves	Crushed leaves are applied on wounds to abate bleeding.
20.	*Emilia sonchifolia* (L.) DC.	Asteraceae	Leaves	Extract of crushed leaves is applied externally on fresh cuts and wounds.
21.	*Euphorbia hirta* L.	Euphorbiaceae	Whole plant	The plant paste is applied over cuts and wounds
22.	*Evolvulus alsinoides* L.	Convolvulaceae	Whole Plant	The plant paste is given orally with milk or hot water to enhance the memory and also it is applied externally to the burn injuries.
23.	*Galinsoga parviflora* Cav.	Asteraceae	Leaves	Plant juice is applied to wounds to check bleeding
24.	*Hedyotis herbacea* L.	Rubiaceae	Whole plant	Whole plant paste is applied over cuts and wounds.
25.	*Impatiens chinensis* L.	Balsaminaceae	Whole plant	The extract of whole plant is used for the application over burns.
26.	*Kalanchoe schweinfurthii* Penzig.	Crassulaceae	Stem and Leaves	The paste obtained from the succulent stem and leaves of the plant is applied over the cuts and wounds.
27.	*Knoxia mollis* Wight & Arn.	Rubiaceae	Leaves	Leaf is ground into a paste and applied topically on affected places to heal wounds.
28.	*Mitracarpus villosus* (Sw.) DC.	Rubiaceae	Leaves	Leaf of this plant and stem bark of *Syzygium cumini* are mixed, ground into a paste and heated with gingelly oil. The mixture thus obtained is applied topically on affected places to heal wounds.
29.	*Opuntia vulgaris* Mill.	Cactaceae	Succulent parts	The succulent part of the plant is crushed and it is applied to inflammation.
30.	*Plantago erosa* Wall.	Plantaginaceae	Leaves	Leaf paste used as an antiseptic in wounds
31.	*Polycarpaea corymbosa* (L.) Lam.	Caryophyllaceae	Whole plant	Poultice made from the fresh leaves is applied over boils and inflammatory swellings.
32.	*Pouzolzia zeylanica* (L.) Bennett.	Urticaceae	Whole plant	Juice extracted by squeezing the leaves is roughed to the right eye with little finger of the left hand three times to cure eye inflammation.
33.	*Pteris confusa* T.G.	Pteridaceae	Rhizome	Rhizome paste is applied over the boils
34.	*Pteris vittata* L.	Pteridaceae	Whole plant	The whole plant parts are ground into paste and it is applied over the affected places for wound healing

(Contd.)

Sl No.	Botanical Name	Family	Parts used	Mode of drug preparation
35.	*Pteridium aquilinum* (L.) Kuhn.	Pteridiaceae	Whole plant	The whole plant parts are ground into paste and applied over the affected places for cuts and wounds
36.	*Pyrrosia porosa* (C. Presl) Hoven.	Polypodiaceae	Whole plant	The whole plant paste is applied over cuts made through knives
37.	*Rubia cordifolia* L.	Rubiaceae	Leaves	Leaf paste is applied over swellings in the gum.
38.	*Rungia repens* (L.) Nees	Acathaceae	Whole plant	Whole plant is ground into a paste and applied topically on affected places to heal wounds.
39.	*Scilla hyacinthina* (Roth) Macbr.	Hyacinthaceae	Bulbs	Bulbs are baked in hot water and tied over the boils and wounds
40.	*Selaginella delicatula* (Desv.) Alston	Selaginellaceae	Whole Plant	Plant juice is antibacterial and it is used for the healing of wounds by tribals.
41.	*Sesamum laciniatum* Klein. *ex* Willd.		Pedaliaceae	Leaves Leaf juice applied on wounds
42.	*Sesamum orientale* L.	Pedaliaceae	Seeds	Seed oil is applied on wounds
43.	*Themeda triandra* Forssk.	Poaceae	Whole plant	Powder of whole plant parts is ground with the leaves of *Toddalia asiatica* and *Pongamia pinnata*. The mixture thus obtained is mixed with coconut oil and applied topically on affected places to treat wounds.
44.	*Tridax procumbens* L	Asteraceae	Leaves	Fresh juice of the leaves is applied to wounds
45.	*Vernonia cinerea* (L.) Less.	Asteraceae	Whole plant	Plant crushed with lime is applied on wounds.

4. CONCLUSION

The present study revealed that traditional medicines are still in common use by the tribal communities. Thus the study ascertains the value of a great number of plants used in tribal medicine especially in wound healing which could be of considerable interest for chemical, pharmacological, clinical and biochemical investigations, which ultimately may birth to drug discovery. The knowledge on such folklore uses of the medicinal plants lead to open up ways for effective utilization of herbal medicines in future. The over exploitation of these plants in the name of medicine may lead some species ultimately to the disappearance in future, therefore special attention should also be made on sustainable utilization of medicinal plants. This study evidently point out that, instead of trying to identify the active components of herbs through massive collection of plants from natural sources, it is better to start investigating the efficacy of the natural product from the traditional use.

REFERENCES

Aravindhan, V., Sathiyadas, K., Rajendran A. and Binu Thomas. 2011. Some rare and endemic medicinal orchids of Velliangiri hills of Southern Western Ghats, Tamil Nadu, India. *Ind. For.* 137 (9): 1077 – 1081.

Binu Thomas, Rajendran, A. and Aravindhan, V. 2012. Chasmophytes: the potential plants for Rock gardening, from Velliangiri Hills of Southern Western Ghats of Tamil Nadu, India. *Bot. Report* 1(1):14 -19.

Binu Thomas, Rajendran, A., Aravindhan, V. and Sivalingam, R. 2012. Ethnobotanical survey of plants used by *Kuruma* tribe for the treatment of cuts and wounds in Kerala, India. *Asian Pac. J. Trop. Biomed.* 1-3.

Biswas, T. K. and Mukherjee, B. 2003. Plant Medicines of Indian Origin for Wound Healing Activity: A Review. *Lower Extr Wounds* 2:25-39.

Gamble, J.S. and C.E.C. Fischer. 1915-1931. *The Flora of the Presidency of madras*. Adlard& Sons Ltd., London. (Repr. Ed. Vols. 1-3. 1957).

Gurib-Fakim, A. 2006. Review – Medicinal plants: Traditions of Yesterday and drugs of tomorrow. *Mol Asp Med* 27:1-93.

Matthew, K.M., (1983). *The Flora of Tamilnadu Carnatic* Parts (1-3)Rapinat Herbarium, Tiruchirapalli.

Mundada S, Shivhare R Pharmacology of *Tridax procumbens* a weed Review. *Int. J Pharm. Tech. Res.* 2010; 2:1391-1394.

Nagori, B.P. and Solanki, R. 2011. Role of medicinal plants in wound healing. *Res. J. Med. Plant* 5(4): 392-405.

Pei, S.J. 2001. Ethno-botanical approaches of traditional medicine studies: Some experiences from Asia. *Pharmaceutical Biol.* 39:74-79.

Sudhakar, M., Rao, G.V. and Selvaraj, J., Senthil Ramanan, R., Radhakrishnan, N. And Murali, M.B. 2003. Efficacy of some indigenous medicines in wound healing in rats. *Ind. J. Animal Sci.* 73:652-656.

19

Anti-inflammatory Activity of *Vitex negundo* L. var. *purpurascens* Sivar. and Mold. leaves Against Carrageenan Induced Oedema on Rats

M. Ramesh and A. Mahalakshmipriya

1. INTRODUCTION

Medicinal plants continue to be an important therapeutic aid for alleviating the ailments of humankind. The search for eternal health and longevity for remedies to relive pain and discomfort drove early man to explore his immediate natural surroundings to the use of many plants, animal products and minerals etc for the development of a variety of therapeutic agents. Today, there is a renewed interest in traditional medicine and an increasing demand for more drugs from plants sources. This revival of interest in plant derived drugs is mainly due to the current widespread belief that "green medicine" is safe and more dependable than the costly synthetic drugs, many of which have adverse side effects.

Vitex negundo or Five-Leaved Chaste Tree is commonly known as Nirgundi in India. It is a large aromatic shrub belonging to the Verbenaceae family of plants. It is distributed throughout the greater part of India up to an altitude of 1500 m in the outer Himalayas. It is a gregarious shrub that is found abundant along the banks or rivers, in moist situations, open waste lands and near the deciduous forests. It is widely planted as a hedge plant along the roads and between the roads. The leaves are tri-or pentafoliate, lateral leaflets are smaller and nearly glabrous. The shrub is one of the important plants used in Indian medicine. Almost all parts of the herb are useful as a drug but the leaves and roots are most important and sold as drugs.

Almost all parts of *V. negundo* are used, the leaves and the barks are the most important in the field of medicine. The decoction of leaves is considered as tonic, vermifuge and is given along with long pepper in cater-rhal fever (Chandramu *et al.*, 2003).

Traditionally it is having the folk claims like useful in treatment of rheumatism, insecticidal, antimicrobial, anticancer, tranquillizer, tonic, febrifuge, expectorant and diuretic properties. The whole plant is said to be a malarial preventative and is also used in the treatment of bacterial dysentery.

Plants have a great potential for producing new drugs for human benefit. Plants used in traditional medicine contain a vast array of substances that can be used to treat chronic and infectious diseases. According to a report of World Health Organization, more than 80% of world's populations depend on traditional medicine for their primary health care needs (Duraipandiyan *et al.*, 2006). The demand for more and more drugs from plant sources is continuously increasing. It is therefore essential for systematic evaluation of plants used in traditional medicine for various ailments. The increased interest in plant derived drugs is mainly because of the wide spread belief that 'herbal medicine' is safer than costly synthetic drugs which possesses side effects. Hence, there is need to screen medicinal plants for promising biological activity. Further, there is a continuous development of resistant strains which pose the need for search and development of new drug to cure diseases (Silver, 1991).

Vitex negundo L. (Verbenaceae) commonly known as nirgundi chiefly occurring throughout India (Watt, 1972; Gupta *et al.*, 2005) is widely distributed in Similipal Biosphere Reserve, Orissa. The plant grows gregariously in wastelands and is also planted as a hedge plant. The bluish purple flower is numerous; the fruit is succulent, black when ripe, rounded and about 4mm in diameter. Though, almost all parts of *V. negundo* are used, the leaves and the barks are the most important in the field of medicine. The decoction of leaves is considered as tonic, vermifuge and is given along with long pepper in catarrhal fever (Chandramu *et al.*, 2003). The genus *Vitex* approximately includes 270 known species of trees and shrubs within tropical and sub tropical regions, although few species may be found in temperate zones. Leaves of this plant have been shown mosquito repellent effects (Hebbalkar *et al.*, 1992) some studies have also been done on antimicrobial activity of *V. negundo* along with some other Indian medicinal plants (Ahmad *et al.*, 1998; Rusia *et al.*, 1998; Kumar *et al.*, 2006; Valsaraj *et al.*, 1997).

An external application of the leaves is considered useful in rheumatic pains, sprains, and inflammations (Kirtikar *et al.*, 1980). The leaves are claimed to possess insecticidal, antitubercular, anticancer properties and the flowers are prescribed in case of fever with vomiting (Ghani, 1998).

2. Materials and Methods

2.1. Plant

The *Vitex negundo L. var. purpurascens Sivar. and Mold.* Plant was collected from the two villeges of Coimbatore (Dist) named Veerappanur and Navakarai , Coimbatore, Tamilnadu. The sample were authenticated by the Botanical Survey of India, Tamil Nadu Agriculture University, Coimbatore, India. The authentication number for the plant is BSI/SC/5/23/09-10/Tech/1327.

2.2. Method of extraction (Soxhalation)

The solvent used in Soxhalation was Methanol and its boiling point is 64.7°C (148.4 °F or 337.8 K). Soxhalation method was used for the extraction of active components from *Vitex negundo* L. *var. purpurascens* Sivar. and Mold. In this method since Soxhlet apparatus was used, it is called Soxhalation. On the laboratory scale, the apparatus consists of a flask, extractor and reflux condenser. The powdered leaves was placed in a thimble and inserted into the wide central tube of the extractor. For each gram of dry material 2 ml of solvent were used. Methanol was used in Soxhlet extractor for 24 hour at a temperature not exceeding the boiling point of the solvent. The Methanolic (MeOH) extract were filtered through Whatman No. 1 filter paper and solvent removed by vacuum distillation in a rotator evaporator at 60°C, care been taken to decrease the temperature to 40°C for final extraction. The residues obtained were designated as Methanolic extract and stored in airtight glass container, protected from sunlight, stored in a freezer at 4°C until assayed.

2.3. Phytochemical analysis

In Preliminary phytochemical study, the extract was screened for the presence of alkaloids, flavonoids, saponins, terpenoids, carbohydrates, glycosides, proteins and amino acids, steroids, vitamin C, fat and fixed oil (Trease *et al.*, 1989).

2.4. Pharmacological activity

2.4.1. Animals

The experiments were performed on albino rats (150-200g) of either sexes (Wister strain) weighing about 150–200g, divided into groups of 6 each. The animals were housed in cages within the facility and allowed free assessment to food and water. Indomethacin (20mg/kg) was used as standard drug. All the animal experiments were approved by the Institutional Animal Ethical Committee (IAEC) and the approval number is KMCRET/Ph.D/07/2012-13

2.4.2. Acute toxicity study

Acute toxicity study was done according to OECD (Organization for Economic Co-operation and Development) 4 Guideline 420. Fixed dose method with starting dose of 2000mg/kg body weight was adopted. The extract were given to a group consist of 6 animals (albino rats) and animals treated groups were kept for observation of behavioral change and death.

2.4.3. Administration of extracts

Doses of 200mg/kg and 400mg/kg of Methanolic extract were prepared in 1ml of distilled water and given orally to separate groups (n=6/group/extract) of rats.

2.4.4. Anti-inflammatory activity

2.4.4.1. Carrageenan induced rat paw oedema (Acute test)

Oedema was induced by subplantar injection of 0.1ml of 1% freshly prepared suspension of carrageenan (Sigma, USA) into the right hind paws of the rats of four groups of six animals each. The volume of the injected right hind paws and left hind paws were measured at 1, 2,3,4,5,6 and 24h and 48h after induction of the inflammation using Digital vernier caliper . The test group received the doses of 200mg/kg and 400mg/kg of methanolic extracts and the standard group received Indomethacin (20mg/kg), and the control rats received the vehicle only. All the treatments were given orally 30 min prior to the injection of carrageenan.

2.5. Statistical analysis

Results obtained were expressed as mean ± SEM. The data were analyzed using the One way ANOVA and results were considered significant when $p< 0.05$, <0.001.

3. RESULTS AND DISCUSSION

In the present study, anti-inflammatory effect of the methanolic leaf extract of *Vitex negundo* L. var. *purpurascens* Sivar. and Mold. was investigated using acute carrageenan model.

Carrageenan-induced hind paw edema is the standard experimental model of acute inflammation. Carrageenan induced edema is a biphasic response. The first phase is mediated through the release of histamine, serotonin and kinins whereas the second phase is related to the release of prostaglandin and slow reacting substances (Vadivelan, 2007).

Table 1: Effect of *Vitex negundo* L. var. *purpurascens* Sivar. and Mold. leaves on carrageenan induced hind paw edema in rats

Treatment and dose (mg/kg ,p.o.)	Time interval (hr) & Edema volume (ml)					
	0h	1h	2h	3h	4h	5h
Control	5.3±0.3	5.6±0.3	5.9±0.1	6.2±0.1	6.4±0.2	6.1±0.2
Indomathacin 20 mg/kg	5.9±0.7	6.1 ±0.3**	6.2 ±0.3*	6.4 ±0.2	6.1 ±0.2*	6.0 ±0.2
Methanolic extracts 200mg/kg	5.6±0.8	5.8±0.5**	6.0±0.1**	6.3±0.6**	6.1±0.1**	5.9±0.6**
Methanolic extracts 400mg/kg	5.4 ±0.6	5.6 ±03**	6.2 ±0.4**	6.5 ±0.3**	6.3 ±0.2**	5.9 ±0.4**

Values of edema are mean ± SEM from 6 animals in each group, while those in parenthesis represent percent inhibition of edema; One-Way ANOVA is applied for statistical analysis. P values: *<0.05; **<0.001.

From the study, the standard (Indomathacin 20 mg/Kg) and treatment groups (methanolic extracts 200mg/Kg & 400 mg/Kg) showed significant reduction in paw volumes from the 4th hour itself when compared with control.(Table:1)

Oral administration of the 400 mg/kg of methanolic extract of *Vitex negundo* L. var. *purpurascens* Sivar. and Mold.showed high significant effect which is near to the level of standard Indomathacin.

4. CONCLUSION

Methanolic extract of *Vitex negundo* L. var. *purpurascens* Sivar. and Mold. has been investigated for phytochemical and pharmacological activities in experimental animals. Phytochemical studies revealed that alcoholic extract of *Vitex negundo* contains Alkaloids, Carbohydrates, cardiac glycoside, Flavonoids, Glycoside, Phenols, Phlobatannins, Proteins, Resins, saponins, Tannins and Terpenoids. In present study,methanolic extract of *Vitex negundo* L. var. *purpurascens* Sivar. and Mold. at dose of 400mg/kg body weight showed significant inhibition of carrageenan induced paw oedema in acute phase.

In conclusion, the study has shown that extract methanolic of *Vitex negundo* L. var. *purpurascens* Sivar. and Mold. possess significant anti-inflammatory effect in laboratory animals at the dose of 400mg/kg. Present investigation together with previous studies provides support to the anti inflammatory properties of *Vitex negundo* L. var. *purpurascens* Sivar. and Mold. leaves. Therefore it can be used as anti inflammatory supplement in the developing countries towards the development of new therapeutic agents. Further pharmacological and clinical studies are required to understand the mechanism and the actual efficiency of these plant extract in treating various infectious diseases.

REFERENCES

Ahmad, I., Mehamood, Z., Mohammad, F., 1998. Screening of some Indian medicinal plants for their antimicrobial properties. *J. Ethnopharmacol.*, 62: 183–193.

Chandramu, C., Manohar, R.D., Krupadanam, D.G.L., Dashavantha, R.V., 2003. Isolation, characterization and biological activity of betulinic acid and ursolic acid from *Vitex negundo* L. *Phytother. Res.* 17: 129- 134

Duraipandiyan, V., Ayyanar, M., Ignacimuthu, S., 2006. Antimicrobial activity of some ethnomedicinal plants used by Paliyar tribe from Tamil Nadu, India. *BMC Comp. Alter. Med.* 6: 35-41.

Ghani, A., 1998. Medicinal plants of Bangladesh, chemical constituents and uses. Dhaka: Asiatic Society of Bangladesh. p. 320.

Gupta, A.K., Tandon, N., Sharma, M., 2005. Quality standards of Indian medicinal plants. Indian Council of Medical Research, New Delhi, India.

Hebbalkar, D.S., Hebbalkar, G.D., Sharma, R.N., Joshi, V.S., Bhat, B.S., 1992. Mosquito repellant activity of oils from *Vitex negundo* Linn leaves. *Ind. J. Med. Res.*, 95: 200-203.

Kumar, V.P., Chauhan, N.S., Padhi, H., Rajani, M., 2006. Search for antibacterial and antifungal agents from selected Indian medicinal plants. *J. Ethnopharmacol.*, 67: 241-245.

Kirtikar, K.R., Bas,u B.D., 1980. Indian medicinal plants, vol. III. 2nd ed India: Lalit Mohan Basu, M.B. 1936-1937.

Rusia, K., Srivastava, S.K., 1998. Antimicrobial activity of some Indian medicinal plants. *Ind. J. Pharm. Sci.*, 60: 57-58.

Silver, L.L., 1993. Discovery and development of new antibiotics: the problem of antibiotic resistance. Antimicrob. Agents. *Chemother.*, 37: 377-383.

Trease, G.E., Evans, W.C., 1989. Text book of Pharmocognosy.13th (eds). Alden Press; Oxford; pp. 512-513.

Valsaraj, R., Pushpangadan, P., Smitt, U.W., Adersen, A., Nyman, U., 1997. Antimicrobial Screening Of Selected Medicinal Plants From India .*J. Ethnopharmacol.*, 58, 75-83.

Watt, G.A., 1972. Dictionary of the Economic Products of India. (Vol. IV & VI), Cosmo Publications, New Delhi, India.

20

Phytochemical Analysis and Antibacterial Activity of *Piper nigrum* Against Food Poisoning Bacteria

R. Gnanasekar, V. Sornajana and K. Umamaheswari

1. INTRODUCTION

Food poisoning is a major concern for both the food industry and consumers inspite of different preservation practices some species of pathogenic bacteria causes food spoilage leading to outbreak of illness. Spices and herbs have been added to food since ancient times, not only as flavouring agents, but also as food medicine and as food preservative (Nakatani, 1994). Additives need for food preservation may be harmful to human health and has led to use of natural products especially plant extracts, their essential in food preservation.

More than 90 percent of the cases of food poisoning each year are caused by *Staphylococcus aureus, Salmonella typhi, Vibrio cholerea, Proteus mirabilis, Pseudomonas aeruginosa, Clostridium perfringens, Campylobacter, Listeria monocytogenes, Vibrio parahaemolyticus, Bacillus cereus* and Entero-pathogenic *Escherichia coli.* These bacteria are commonly found on many raw foods and cause severe disorder in humans. Spice posses anti adhesive properties that prevent the microbe to the host tissue, thus preventing the primary infection. It also possesses a rich array of phytochemical and a rich source of antioxidant. Some studies claim that the phenolic compounds present in spices and herbs might also play a major role in their antimicrobial effects.

Piper nigrum finds extensive use in the ayurvedic system of medicine. It possesses CNS stimulation, antipyretic, analgesic and anti feedant activities. Pepper is a natural spice. It is cultivated throughout the world and its usage as a spice is well known. Active components of spices / herbs at low concentrations may interact synergistically with other factors (NaCl, acids, and preservatives) to increase preservative effect. Nutrients present in spices /herbs may stimulate growth and/ or biochemical activities of microorganisms. Thus, food product safety and shelf life depend in some part on the type, quantity, and character of spices and herbs added to the products (Zaika, 1988). The present study was aimed to determine the antibacterial properties of *Piper nigrum.*

2. MATERIALS AND METHODS

2.1. Spice Material

The spice material both black and white (*Piper nigrum*) was collected from the local market and authenticated. The spices were powdered and the alcoholic (ethanol) extracts were prepared by soxhlet apparatus (Jeevan *et al.*, 2004). 25 grams of powdered sample were put in a soxhlet apparatus with 100ml of the solvent. The resulting extract was filtered, concentrated, dried and used for further analysis.

2.2. Test Organism

The common food poisoning bacterial species were used as test organism. The species included *E.coli, Staphylococcus aureus, Pseudomonas aeruginosa, Vibrio cholerea, Salmonella typhi and Proteus mirabilis.* The inhibitory activity of alcoholic extract on the bacterial species was determined by tube dilution technique.

2.3. Determination of MIC and MBC

The alcoholic extracts of black and white pepper were serially diluted in sterile Mueller Hinton broth taken in test tubes to obtain a concentration ranging from 32mg/ml to 0.125mg/ml. 20μl of the bacterial suspension whose turbidity was matched with 0.5 Mc Farland standard was added to each of the test tube and were incubated at 37°C for 24 hrs. The minimum dilution of the extract that suppressed visible growth of the bacterial species was recorded as MIC. The minimum bactericidal concentration (MBC) was done by plating each of the dilution on to sterile Mueller Hinton agar plates and the minimum dilution that killed the isolates was recorded.

2.4. Phytochemical analysis

The presence of phytochemicalssuch as tannins, flavonoids, alkaloids, saponins and terpenoids in the ethanolic extracts were qualitatively analysed (Edeoga *et al.*, 2005).

3. RESULTS AND DISCUSSION

The ethanolic extracts of the unripened and ripened dried fruits of *Piper nigrum* were prepared using soxhlet apparatus. The extracts were tested against the most common food spoilage bacteria including *E.coli, Staphylococcus aureus, Pseudomonas aeruginosa, Vibrio cholerea, Salmonella typhi* and *Proteus mirabilis.*

The antibacterial activity of the ethanolic extracts was tested and the minimum inhibitory concentration and minimum bactericidal concentration were determined by macro broth tube dilution method The Mueller- Hinton broth was used for the testing. The activities of the extract against the microorganism were noted.

The ethanolic extracts of the black pepper showed an inhibitory activity at 1mg/ml against *E.coli*, 4mg/ml against *Staphylococcus aureus* and *Psuedomonas aeruginosa*, 8mg/ml against *Vibrio cholerea*, *Salmonella typhi* and Proteus *mirabilis*. The MBC value of ethanolic extract of black pepper ranged from 4mg/ml to 16mg/ml (Table 1)

Table 1: Minimum inhibition concentration and minimum bacterial concentration of ethanolic extract of Black pepper

S.No.	Microorganism tested	MIC (mg/ml)	MBC (mg/ml)
1.	*Escherichia coli*	1	4
2.	*Staphylococcus aureus*	4	8
3.	*Proteus mirabilis*	8	16
4.	*Salmonella typhi*	8	16
5.	*Pseudomonas aeruginosa*	4	8
6.	*Vibrio cholerea*	8	16

The ethanolic extracts of the white pepper showed an inhibitory activity at 4mg/ml against E.coli, *Psuedomonas aeruginosa,* 2mg/ml against *Staphylococcus aureus*, 8mg/ml against *Proteus mirabilis,* while *Salmonella typhi* was not inhibited.

The MBC value of ethanolic extracts of the white pepper was found to be 8mg/ml against *E.coli,* 4mg/ml against *Staphylococcus aureus*, 16mg/ml against *Proteus mirabilis,* 8mg/ml against *Psuedomonas aeruginosa,* 16mg/ml against *Vibrio cholerea* and no sensitive against *Salmonella typhi* (Table 2).

Table 2: Minimum inhibition concentration and minimum bacterial concentration of ethanolic extract of White pepper

S.No.	Microorganism tested	MIC (mg/ml)	MBC (mg/ml)
1.	*Escherichia coli*	4	8
2.	*Staphylococcus aureus*	2	4
3.	*Proteus mirabilis*	8	16
4.	*Salmonella typhi*	No activity	No activity
5.	*Pseudomonas aeruginosa*	4	8
6.	*Vibrio cholerea*	8	16

The extracts were tested for its phytochemical constituents as per standard protocol. The ethanolic extracts of the black pepper and white pepper showed positive reaction for the presence of saponins, tannins, alkaloids, terpenoids, flavonoids and steroids and cardiac glycosides were absent in the extracts tested (Table 3).

Table 3: Phytochemical constituents of Black Pepper and White pepper

S.No.	*P. nigrum*	Solvent	Name of phytochemicals							
			STF	A	S	T	P	CG	TE	
1.	Black pepper	Ethanol	_	++	+	+	+	++	_	+
2.	White pepper	Ethanol	_	+	+	+	+	+	_	+

ST: Steroids, F: Flavonoids, A: Alkaloids, S: Saponins, T: Tannins, CG: Cardiac glycoside, TE: Terpenoids

4. CONCLUSION

The increasing antibiotic resistance of some pathogens that are associated with food borne illness is major concern (Meng *et al.*, 1998; Perreten *et al.*, 1998; Stermitz *et al.*, 2000). The interest in the development of new types of effective and nontoxic antimicrobial compounds, has led to the use of natural antibacterial compounds, such as extracts of spices and herbs, for food preservation (Smid and Gorris, 1999). In addition to imparting characteristic flavors, certain spices are valued for their antimicrobial activities and medicinal effects in addition to their flavor and fragrance qualities. Plants including spices and herbs contain complex phenolics (e.g., phenolic acids, flavonoids, tannins, lignans, coumarins, quinines) that contribute to their antimicrobial property.

In the present study, *Piper nigrum* which has been widely used as a spice has been evaluated for the antibacterial properties against some of the most common food poisoning bacteria such as *E.coli, Staphylococcus aureus, Pseudomonas aeruginosa, Vibrio cholerea, Salmonella typhi* and *Proteus mirabilis.* The dried unripened fruit and the dried and coat removed ripened fruit of *Piper nigrum* commonly referred as black and white pepper respectively were used in the present study, the six species of bacteria tested were inhibited at MIC value ranging from 1-8 mg/ml, and *E.coli* was inhibited at a less concentration when compared to *Proteus mirabilis*. Shan *et al.,*2007 has reported the activity of methanolic extracts of *Piper nigrum* against these bacterial strains found the extracts, were inactive against the bacterial species tested while in the present study we have evaluated the activity of extracts prepared in ethanol and the extracts were found to be more active.

The inhibitory activity of both black and white pepper were comparable. The MIC values of the two active spice extracts obtained in this study were lower than the MBC values suggesting that the plant extracts were bacteriostatic at lower concentration and bactericidal at higher concentration. The ethanol extract of two spice exerted greater antibacterial activity at the same concentrations.

The qualitative analysis of the phytochemicals revealed the presence of tannins, terpenoids, flavanoids and alkaloids and cardiac glycosides, steroids are not detected. The presence of phenol and other phytochemicals may have contributed to the inhibitory activity of the organism tested.

The stronger extraction capacity of ethanol could have produced greater number of active constituents responsible for antibacterial activity. Further analysis of the phytochemicals and quantitative determination and testing the individuals compounds may provide a broader insight to the antimicrobial properties of *Piper nigrum*.

REFERENCES

Jeevan Ram, A., Bhakshu, L.M., Venkata Raju, R.R., 2004. *In vitro* antimicrobial activity of certain medicinal plants from Eastern Ghats, India, used for skin diseases. *J Ethanopharmacol*. 90(2-3): 353-7.

Meng, J.H., Zhao, S.H., Doyle, M.P., Joseph, S.W., 1998. Antibiotic resistance of *Escherichia coli* O157: H7 and O157: NM isolated from animals, food, and humans. *J Food Prot.* 61(11): 1511-4.

Perreten, V., Giampa, N., Schuler-Schmid, U., Teuber, M., 1998. Antibiotic resistance genes in coagulase-negative *Staphylococci* isolated from food. *Syst Appl Microbiol.* 21(1): 113-20.

Smid, E.J., Gorris, L.G.M., 1999. Natural antimicrobials for food preservation. In: Rahman, M.S. (Ed.), Handbook of Food Preservation. Marcel Dekker, New York, pp. 285–308.

Stermitz, F.R., Tawara-Matsuda, J., Lorenz, P., Mueller, P., Zenewicz, L., Lewis, K., 2000. 52 -Methoxyhydnocarpin-D and pheophorbide A: Berberis species components that potentiate berberine growth inhibition of resistant *Staphylococcus aureus*. *J Nat Prod.* 63(8): 1146-9.

Zaika, L.L., 1988. Spices and herbs: Their antimicrobial activity and its determination. *J Food Safety.* 9: 97-118.

Mbaebie, B.O., Edeoga, H.O., Afolayan, A.J., 2012. Phytochemical analysis and antioxidants activities of aqueous stem bark extract of *Schotia latifolia* Jacq. *Asian Pacific Journal of Tropical Biomedicine.* 9: 118-124.

21

Evaluation of Anti-diabetic Potential of *Phyllanthus emblica* L.

Renisheya Joy Jeba Malar, T., Kalaiarasi, V. and Johnson, M.

1. INTRODUCTION

Medicinal plants are the most important source of life saving drugs for the majority of the World's population. Plants have been an important source of medicine for thousands of years. Diabetes is one of the major degenerative diseases in the world today. It is a major risk factor for the development of cardiovascular disease. About 70 - 80% of deaths in diabetic patients are due to vascular disease. Diabetes mellitus is also associated with hyperlipidaemia with profound alteration in the concentration and composition of lipid (Odetola *et al.*, 2006). Changes in the concentration of lipid with diabetes contribute to the development of vascular disease. Excessive levels of blood cholesterol accelerate atherogenesis and lowering high blood cholesterol reduces the incidence of CHD Grundy (1986). One of the risk factors for coronary heart disease is elevated Total Cholesterol (TC), Low Density Lipoprotein-Cholesterol (LDL-C) and Lower High Density Lipoprotein-Cholesterol (HDL-C) (Hu *et al.*, 2001; Schaefer, 2002; Moreno and Mitjavalia, 2003; Jain *et al.*, 2007). In particular, hyperglycemia, the primary clinical manifestation of diabetes, is thought to contribute to diabetic complications by altering vascular cellular metabolism, vascular matrix molecules and circulating lipoproteins (Inoguchi *et al.*, 1994). The high level of glucose in the blood, erythrocytes has become important in the pathophysiologic mechanisms in diabetics as shown from several abnormal features demonstrated for red cells of diabetes mellitus patients (Adisa *et al.*, 2004). In traditional medicine, diabetes mellitus is treated with diet, physical exercise and medicinal plants. Ethnobotanical information indicates that more than 800 plants are used as traditional remedies for the treatment of diabetes (Pushparaj *et al.*, 2000).

The genus *Phyllanthus* (Euphorbiaceae) contains 550-750 species. It is common all over tropical and sub-tropical India and also found in Burma. *Phyllanthus emblica* L. commonly known as amla is an ancient herb used in folk medicine for the treatment of variety of human ailments particularly diabetes. Amla has been considered the best of the Ayurvedic rejuvenative herbs. Uniquely, it has a natural balance of tastes (sweet, sour, pungent, bitter and astringent) all in one fruit, it stimulates the brain to rebalance the three main components of all physiological functions, the water, fire, and air elements within the body (Bajracharya, 1979). The fruit is a very rich source of vitamin C and every 100g of fresh fruit provides 470-680 mg of vitamin C. The vitamin value of amla increased further when the juice was extracted from the fruit. The dehydrated berry provided 2428-3470 mg of vitamin C per 100 g. Amla fruits are used in the treatment of diabetes (Nadkarni and Nadkarni, 1999). Decoctions of the leaves and seeds are used in the treatment of diabetes mellitus (Treadway, 1994). Aqueous amla fruit extract has shown significant decrease of the blood glucose as well as triglyceridemic levels and an improvement of the liver function caused by a normalization of the liver-specific enzyme alanine transaminase (ALT) activity (Qureshi *et al.*, 2009). With this knowledge, the present study was intended to reveal the anti-diabetic potential of *P. emblica.*

2. MATERIALS AND METHODS

2.1. Collection of Plant Materials

The whole plants of *Phyllanthus emblica* L. was collected from Nagercoil, Kanyakumari district, Tamil Nadu, India. The fresh materials were shade dried and powdered using mechanical homogenizer. 25 g air dried powder of the plant sample was extracted by cold extraction with 150 mL of methanol for 72 h under dark with intermittent shaking. After incubation, the solution was filtered through filter paper and the filtrate was collected and stored at 4°C for the further studies (crude extracts).

2.2. Antidiabetic activity

2.2.1. Non-enzymatic haemoglycosylation method

The anti-diabetic activities of the methanolic extracts of *P. emblica* wasdetermined by the method of O-toluidine using the modified reagent (Sasaki *et al.*, 1972).

2.2.2. LDL Oxidation

Measurement of LDL oxidation in the plant extract was analysed using the iodometric assay designated by Girotti *et al.*, 1985 and the LDL oxidation was estimated using fluorescence detection (Jessup *et al.*, 1990).

3. RESULTS AND DISCUSSION

3.1. Haemoglobin glycosylation

The anti-diabetic activity of *P. emblica* methanolic extracts was carried out with various concentrations (1, 5, 10, 20, 50 and 100 µg) by Haemoglobin glycosylation test and the results were tabulated in table 1. Highest haemoglobin glycosylation activity 62.51 % was observed in 100 µg of *P. emblica* methanolic extract and the lowest inhibition activity perceived in 1 µg (6.42%).

Table 1: Inhibitory effects of the methanolic extract of *P.emblica* on Haemoglobin glycosylation

Sample Concentration	% Haemoglobin glycosylation
1µg	6.42%
5 µg	12.95%
10 µg	20.10%
20 µg	19.97%
50 µg	43.09%
100 µg	62.51%

3.2. LDL Oxidation

The anti-diabetic activity of *P. emblica* methanolic extract was determined by LDL Oxidation method with various concentrations viz., 0.025, 0.05, 1, 2, 5 and 10 µg and the estimates were recorded in Table 2. 5 µg methanolic extract of *P. emblica* showed the highest LDL Oxidation activity (0.6089) and 0.025 µg exposed the lowest LDL oxidation activity (0.316).

Table 2: Changes in the absorbance at 234nm with $CuSO_4$ induced LDL Oxidation of methanolic extract of *P. emblica*

Sample Concentration	OD @				
	0 min	30 min	60 min	120 min	180 min
0.025 µg	0.316	0.278	0.267	0.234	0.230
0.05 µg	0.768	0.227	0.195	0.104	0.095
1 µg	0.3231	0.5232	0.469	0.473	0.286
2 µg	0.3784	0.483	0.478	0.479	0.407
5 µg	0.6089	0.374	-ve	-ve	-ve
10 µg	0.674	0.316	-ve	-ve	-ve
Control	0.4846	0.7730	0.6449	0.4105	0.2137

The methanolic extract of *P. emblica* were displayed the various degree of anti-diabetic activity. Similar to our observations, Ashwanth Ram *et al.*, (2008) observed the total antidiabetic capacity assay; 1 mg of extract is equivalent to 41.31 μg of ascorbic acid. It showed 83.69% inhibition of haemoglobin glycosylation in *Phyllanthus reticulatus*. James *et al.*, (2011) reported methanolic leaf extract inhibit haemoglobin glycosylation, where as concentration of 20 mg/ml of the extract gave significant inhibition by yielding haemoglobin concentration of 1.877±0.40 μg/ml for test extract against 0.032±0.013 μg/ml for the standard. But in the present study we observed 62.51% of haemoglobin glycosylation at 100 μg/ml of methnaolic extract of *P. emblica*. Control of diabetes mellitus normally involves exercise, diet and chemotherapy. Diabetes can be treated by antioxidants due to their various therapeutic benefits. They include phytochemicals, vitamins and other nutrients that protect our cells from damage caused by free radicals. Antioxidants can fit in two broad categories – enzymatic and non-enzymatic. They provide the necessary defense against the OS generated by ROS. It has an array of non-enzymatic antioxidants-ascorbate, urate (Thiele *et al.*, 1995), vitamin - E (Aitken and Clarkson, 1998), pyruvate (De lamirande and Gagnon, 1992), glutathione (Lenzi *et al.*, 1994), albumin, vitamin-A, ubiquinol (Tanber *et al.*, 1975), taurine and hypotaurine (Alvarez and Storey, 1983). Most of the fruits, vegetables, culinary herbs and medicinal herbs contain high levels of antioxidants (Dragland, 2003). A botanical diversity of fruits and vegetables plays a role in the biological effect of antioxidant phytochemicals and foods that contain antioxidants in our daily diet (Radhika *et al.*, 2011). Khopde *et al.*, (2001) and Charoenteeraboon *et al.*, (2010) confirmed the antioxidant potential of *P. emblica*. The previous antioxidant observations were used as source for present study and supported the present study findings also. The methnaolic extracts of *P. emblica* showed 62.51% of haemoglobin glycosylation at 100 μg/ml. High percentage haemoglobin glycosylation potential of *P. emblica* may be due to the presence antioxidant properties of *P. emblica*. During diabetes, the excess glucose present in the blood reacts with haemoglobin to form glycosylated haemoglobin. It has been reported that various proteins, including haemoglobin, albumin, collagen, LDL, or crystalline proteins undergo non-enzymatic glycation in diabetes (Klein, 1995). The rate of glycation is proportional to the concentration of blood glucose. Glycosylated haemoglobin has been found to be increased over a long period of time in diabetes (Forbes *et al.*, 2003). There is evidence that glycation itself may induce the formation of oxygenderived free radicals in diabetic condition (Gupta *et al.*, 1997). Therefore, the measurement of glycosylated haemoglobin is supposed to be a very sensitive index for glycemic control. In the present study also we measured percentage of haemoglobin glycosylation; we observed high percentage of haemoglobin glycosylation with low volume of methanolic extracts. The present study demonstrates that methnaolic extracts of *P. emblica* has significant anti-diabetic potential.

4. CONCLUSION

Further pharmacological properties can be analyzed in the future that will help to predict the novel drugs from *P. emblica* to treat diabetics and related heart diseases with less side effects.

REFERENCES

Adisa, R.A., Oke, J., Olomu, S.A., Olorunsogo, O., 2004. Inhibition of human haemoglobin glycosylation by flavonoid containing leaf extract of *Cnestis ferruginea*. *Journal of the Cameroon Academy of Sciences* 4: 351-359.

Aitken, R.J., Clarkson, J.S., 1998. Significance of reactive oxygen species and antioxidants in defining the efficacy of sperm preparation techniques. *J. Androl.*, 9: 367.

Alvarez, J.G., Storey, B.T., 1983. Taurine, hypotaurine, epinephrine 1 and albumin inhibit lipid peroxidation in rabbit spermatozoa and protect against loss of motility. *Biol. Reprod.*, 29:548.

Aswatha Ram, H.N., Shreedhara, C.S., Falguni, P., Gajera, Sachin, B., Zanwar, 2008. *In vitro* Free Radical Scavenging Potential of Methanol Extract of Entire Plant of *Phyllanthus reticulates. Poir Pharmacologyonline,* 2: 440-451.

Bajracharya, M.B., 1979. Ayurvedic Medicinal Plants. Kathmandu; Piyusavarsi Ausadhalaya.

Charoenteeraboon, J., Ngamkitidechakul, C., Soonthornchareonnon, N., Jaijoy, K., Sireeratawong , S., 2010. Antioxidant activities of the standardized water extract from fruit of *Phyllanthus emblica* Linn. Songklanakarin. *J. Sci. Technol.* 32(6): 599-604.

De lamirande, E., Gagnon, C., 1992. Reactive oxygen species and human spermatozoa. II. Depletion of adenosine triphosphate plays an important role in the inhibition of sperm motility. *Androl*, 13: 379.

Dragland, S., 2003. "Several Culinary and Medicinal Herbs are Important Sources of DietaryAntioxidants", *J. Nutrition*.

Forbes, M., Cooper, M.E., Oldfield, M.D., Thomas, M.C., 2003. Role of advanced glycation end products in diabetic nephropathy. *J. Am Soc. Nephrol*, 14: S254-S258.

Grundy, S.M., 1986. Comparison of mono unsaturated fatty acid and carbohydrates for lowering plasma cholesterol. *N. Eng. J. Med.*, 314: 745-748.

Gupta, B.L., Nehal, M., Baquer, N.Z., 1997. Effect of experimental diabeteon the activities of hexokinase, glucose-6-phosphate dehydrogenas and cathechoamines in rat erythrocytes of different ages. *Indian Exp. Biol.*, 35: 792-795.

Girotti, A.W., Thomas, J.P., Jordan, J.E., 1985. Lipid photooxidation in erythrocyte ghosts: sensitization of the membranes toward ascorbate- and superoxide-induced peroxidation and lysis. Arch. Biochem. *Biophys.*, 236: 238-251.

Hu, F.B., Manson, J.E., Willett, W.C., 2001. Types of dietary fat and risk of coronary heart disease: A critical review. *J. Am. Coll. Nutr.*, 19: 5.

Inoguchi, T., Xia, P., Kunisaki, M., Higashi, S., Feener, E.P., King, G.L., 1994. Insulin's effect on proteinkinase C and diacylglycerol induced by diabetes and glucose in vascular tissues. *American Journal of Physiology*, 267: E369- E379.

James, S.A., Auta, R., Goje, D.J., 2011. *In vitro* study on inhibition of glycosylation of methanolic leaf extract of *hibiscus cannabinus. Science World Journal,* 6(3): 7-9.

Jain, K.S., Kathiravan, M.K., Somani, R.S., Shishoo, C.J., 2007. The biology and chemistry of hyperlipidemia. *Bioorg. Med. Chem.*, 15: 4674.

Jessup, W., Rankin, S.M., De Whalley, C.V., Hoult, J.R.S., Scott, J., Leak, D.S., 1990. α-Tocopherol consumption during low-density-lipoprotein oxidation. *Journal of Chemical and Pharmaceutical Research*. 3(2): 204-212.

Klein, R., 1995. Hyperglycemia and microvascular and macrovascular disease in diabetes. *Diabetes Care*, 18: 258-268.

Khopde, S.M., Indira Priyadarsini, K., Mohan, H., Gawandi, V.B., Satav, J.G., Yakhmi, J.V., Banavaliker, M.M., Biyani, M.K., Mittal, J.P., 2001. Characterizing the antioxidant activity of amla (*Phyllanthus emblica*) extract. *Current science*, 81(2): 185-190.

Lenzi, A., Picardo, M., Gandhini, L., 1994. Glutathione treatment of dyspermia: Effect on the lipid peroxidation process, *Human Reproduction*, 9: 2044.

Moreno, J.J., Mitjavalia, M.T., 2003. The degree of unsaturation of dietary fatty acids and the development of atheroscelerosis. *J. Nutr. Biochem.*, 14: 182.

Nadkarni, K.M., Nadkarni, A.K., 1999. Indian Materia Medica - with Ayurvedic, Unani-Tibbi, Siddha, Allopathic, Homeopathic, Naturopathic and Home remedies. Vol.1. Popular Prakashan Private Ltd., Bombay, India.

Odetola, A.A., Akinloye,O., Egujobi,C., Adenkunle W.A., Ayoola, A,O., 2006. Possible antidiabetic and antihyperlipidemic effect of fermented *Parkia Biglobosa* (JACQ) extract in alloxan-induced diabetic rats. *Clin. Exp. Pharmacol. Physiol.*, 33: 808-812.

Pushparaj, P., Tan, C.H., Tan, B.K.H., 2000. Effects of *Averrhoa bilimbi* leaf extact on blood glucose and lipids in streptozotocin-diabetic rats. *J. Ethnopharmacol.*, 72: 69-76.

Qureshi, S.A., Asad, W., Sultana, V., 2009. "The Effect of *Phyllantus emblica* Linn on Type – II Diabetes, Triglycerides and Liver – Specific Enzyme"]. *Pakistan Journal of Nutrition*. 8(2): 125–128.

Radhika, S., Smila, K,H., Muthezhilan, R.,2011. Antidiabetic and hypolipidemic activity of *Punica granatum* Linn on alloxan induced rats. *World Journal of Medical Sciences*, 6(4): 178-182.

Sasaki, T., Matsy, S, Sorae, A., 1972. Effect of acetic acid concentration on the colour reaction in the Otoluidine boric acid method for blood glucose estimation. *Rinsho Kagarku*, 1:346-353.

Thiele, J.J., Friesleben, H.J., Fuchs, J., 1995. Ascorbic acid and urate in Human seminal plasma: Determination and interrelationships with chemiluminescence's in washed semen. *Human Reproduction*, 10: 10.

Treadway, Linda, 1994. Amla Traditional food and medicine. HerbalGram. *The Journal of the American Botanical Council.* 31: 26.

22

Nutraceutical and Phytotherapeutical Properties of Buckwheat: A Review Update

Thankarajan Sajeesh and Thangaraj Parimelazhagan

1. INTRODUCTION

Buckwheat is one of the most important life supporting nutritious crops, which has come as a secondary crop in many countries and has now attained its position in the agriculture of many countries where cereals form the stable food for the people. Though the crop is not a cereal, the seeds themselves can bring the name 'pseudocereal' due to their similar usage as cereal grains. There are many species of buckwheat in the world, of which mainly nine species have agricultural value (Krkoskova and Mrazova, 2005). Generally, the genus *Fagopyrum* in the family Polygonaceae includes two buckwheat species used for food around the world: common buckwheat (*F. esculentum* Moench), which is almost grown on all continents, and tartary buckwheat (*F. tataricum* (L.) Gaertn), which originated in eastern Tibet or northwestern Yunnan in China and is grown only in Asia, Europe, and North America (Bonafaccia *et al.*, 2003; Tsuji and Ohnishi, 2001).

Buckwheat is grown throughout a large area as a crop that fits the farming system on marginal and fairly unproductive land. It is used as a subsistence crop in many of the more mountainous areas where it is often grown with barley at the higher altitudes. Tartary buckwheat, because of its frost tolerance, is generally grown at the higher altitudes whereas common buckwheat is grown at the lower altitudes. In many areas the trend is for replacement of common buckwheat, which has lower yielding ability and lacks frost tolerance, with finger millet or other crops.

Tartary buckwheat production in most areas appears to be remaining constant (Campbell, 1997).

The grain is generally used as human food and as animal or poultry feed, with the dehulled groats being cooked as porridge and the flour used in the preparation of pancakes, biscuits, noodles, cereals, etc (Campbell, 1997). However, it is being introduced in many countries because buckwheat seed is a nutritionally important food and has beneficial effects on human health, due to its composition. The protein of buckwheat is of excellent quality, unlike cereals which are deficient in lysine, one of the essential amino acids for human health (Campbell, 1997; Watanabe, 1998). It is also a multipurpose crop. The tender leaves and shoots are used as leafy vegetable, the flowers and green leaves are used for extraction of rutin, which is used in medicine (McGregor and McKillican, 1952), and the flower produces honey of very good quality (Campbell, 1997). The seed is used in a number of culinary preparations as well as alcoholic drinks; the husk is used in stuffing pillows. The crop helps in soil binding and checks soil erosion during the rainy seasons (Joshi and Paroda, 1991). It is also a good green manure crop and improves soil texture (Ladefoged, 1952).

2. PHYTOCHEMICAL STUDIES

Primary metabolites

The main consumption buckwheat form is seed. It is a rich source of starch and contains many valuable compounds, such as proteins, antioxidative substances, trace elements and dietary fibre (Bonafaccia *et al.*, 2003; Kreft *et al.*, 1999; Park, 2000). Table 1 shows the nutritional profile of common and tartary buckwheat compared with the common flours and different processed parts. Whole buckwheat groats contain 55% starch, 12% protein, 4% lipid, 2% soluble carbohydrates, 7% total dietary fibre, 2% ash and 18% other components (organic acids, phenolic compounds, tannins, phosphorilated sugars, nucleotides and nucleic acids and unknown compounds). Buckwheat bran is a rich source of total dietary fibre and soluble dietary fibre, particularly bran with hull fragments (40% of which 25% is soluble dietary fibre), while bran without hull fragments has 16% total dietary fibre of which 75% is soluble dietary fibre (Bonafaccia *et al.*, 2003; Steadman, 2001). With typically 80% unsaturated fatty acids and more than 40% polyunsaturated essential fatty acid–linoleic acid, buckwheat is nutritionally superior in fatty acid composition to cereal grains (Bonafaccia *et al.*, 2003; Steadman, 2001; Bonafaccia and Fabjan, 2003; Qin, 2011).

The amino acid composition of common and tartary buckwheat parts are given in table 2. The seeds and seed sprouts of common buckwheat are rich in nutrients and phenolic compounds, and show a good balance of amino acids and minerals (Steadman, 2001; Kim, 2004; Pomeranz and Robbins, 1972). Buckwheat grains form an excellent supplement for cereal grains because it contains large amounts of protein, starch and vitamins (Maeda, 2004).

Table 1: Chemical composition of some cereal flours, *F. esculentum* and *F. tataricum* (g/100 g DW ± SD)

	Protein	Ash	Fat	Starch	Dietary fibre
Wheat[a]	11.5	1.7	1.0	-	2.4
Oats[a]	12.6	1.8	7.1	-	10.2
Rye[a]	11.7	1.5	1.8	-	13.6
Spelt wheat[a]	13.5	1.9	2.5	-	6.0
F. esculentum					
Grain[b]	11.7±0.04	2.19±0.01	2.88±0.06	55.8±0.15	27.38±0.19
Bran[b]	21.6±0.08	4.08±0.01	7.20±0.04	40.7±0.21	26.37±0.27
Flour[b]	10.6±0.04	1.82±0.02	2.34±0.08	78.4±27	6.77±0.18
Raw whole seeds[c]	15.46±0.18	—	3.82±0.16	62.46±0.35	—
Soaked whole seeds[c]	15.29±0.11	—	3.80±0.13	60.08±0.33	—
Streamed whole seeds[c]	13.12±0.19	—	2.75±0.08	59.33±0.29	—
Dried whole seeds[c]	13.09±0.05	—	2.64±0.17	57.96±0.08	—
Dehulled groats[c]	6.23±0.05	—	0.67±0.04	70.36±0.32	—
Buckwheat tea[c]	6.02±0.04	—	0.50±0.08	68.28±0.13	—
F. tataricum					
Grain[b]	11.1±0.01	2.81±0.01	2.81±0.09	57.4±0.12	25.97±0.35
Bran[b]	25.3±0.06	4.97±0.01	7.35±0.11	37.6±0.31	24.76±0.26
Flour[b]	10.3±0.02	1.80±0.01	2.45±0.08	79.4±0.29	6.29±0.12

[a] Bonafaccia and Fabjan, 2003; [b] Bonafaccia *et al.*, 2003; [c] Qin, 2011

Guo and Yao (2006) extracted protein fractions (albumin, globulin, prolamin and glutelin) from defatted tartary buckwheat flour. Albumin was found to be the predominant protein fraction (43.8%) followed by glutelin (14.6%), prolamin (10.5%), and globulin (7.82%). The buckwheat protein is widely reported to have many unique physiological functions, such as curing chronic human diseases, decreasing blood cholesterin (Kayashita *et al.*, 1995; Tomotake *et al.*, 2000), inhibiting mammary cancer caused by 7,12-dimethylbenzene restraining gall-stone (Tomotake *et al.*, 2000; Kayashita *et al.*, 1999; Guo and Yao, 2005).

The seed protein contents of ten species of genus *Fagopyrum* were determined by means of the Kjeldahl method by Guo *et al.* (2007). The average protein content of *F. esculentum* was found to be 12.94%, whereas that of tartary buckwheat was 12.17%. However, it was observed that there are significant differences of seed protein content among different buckwheat species and between different buckwheat types such as between the diploid and the tetraploid, between the cultivated and the wild, and between the big-achene group and the small-achene group. It is generally recognized that albumin and globulin are the major storage proteins in buckwheat seeds, and prolamin and glutelin contents are very low (Guo and Yao, 2006). The major storage proteins from common buckwheat seeds, including 8 S and 13 S globulins, and 2 S albumins have been also characterized (Fujino *et al.*, 2001; Milisavljevic *et al.*, 2004).

Table 2: Essential Amino acid composition of Buckwheat and its sprouts

	F. esculentum[a] (g/100 g protein)		*F. tartaricum*[a] (g/100 g protein)		Buckwheat Seed[b] (mg/100 g DW)	Buckwheat sprouts[b] (mg/100 g DW)		
	Bran	Flour	Bran	Flour		3 DAS	5DAS	7DAS
Valine	5.13	5.23	5.19	5.19	26.3±2.3	40.6±3.5	64.0±4.5	119.7±7.9
Leucine	6.51	6.92	6.35	7.11	6.8±1.4	22.7±1.9	38.4±3.0	74.9±3.9
Isoleucine	3.77	3.93	3.96	4.23	9.6±1.5	14.6±2.1	31.1±2.0	68.3±4.1
Methionine	1.09	1.41	1.33	1.42	3.7±0.1	7.5±1.0	8.8±2.2	14.7±3.0
Phenylalanine	4.54	4.62	4.46	4.71	3.7±0.3	17.6±2.1	26.6±3.3	68.6±3.2
Threonine	3.55	3.71	3.47	3.72	12.4±1.7	26.7±1.9	46.7±3.6	94.9±7.6
Lysine	5.47	5.84	5.88	6.18	7.0±1.1	22.0±0.5	41.4±3.2	56.2±2.6
Arginine	10.5	9.91	11.0	9.63	73.0±5.2	115.6±11.0	224.5±12.1	146.0±7.8
Histidine	2.66	2.47	2.73	2.62	3.7±0.3	19.1±1.1	56.2±3.6	144.3±10.2

[a]Bonafaccia *et al.*, 2003; [b]Kim *et al.*, 2004

The contents of protein, lipid, starch, dietary fibre and vitamins B1, B2 and B6 were analysed in the flour and bran of both common and tartary buckwheat. In both species most lipid substances are concentrated in the bran and there was a prevalence of unsaturated fatty acids—C18:1, C18:2, C18:3 and C20:1. Table 3 depicts that there were relatively small differences in the contents of vitamins B1 and B2 between the two main utilisable milling fractions, but more substantial differences in the contents of vitamins B6 (up to 0.61 mg/100 g in the tartary buckwheat bran fraction). However, total B vitamin content was higher in tartary buckwheat than in common buckwheat (Bonafaccia *et al.*, 2003).

Table 3: Vitamin contents in common and tartary buckwheat (mg/100 g)

	Vitamin B1	Vitamin B2	Vitamin B6
Common buckwheat			
Grain	0.22±0.06	0.10±0.03	0.17±0.04
Bran	0.31±0.08	0.21±0.03	0.58±0.06
Flour	0.28±0.02	0.14±0.05	0.15±0.04
Tartary buckwheat			
Grain	0.41±0.04	0.12±0.02	0.25±0.03
Bran	0.61±0.08	0.32±0.04	0.61±0.06
Flour	0.40±0.07	0.28±0.02	0.18±0.03

The contents of Se, Cr, Rb, Zn, Fe, Co, Sb, Ba, Ni, Ag, Hg and Sn in the flour and bran of common and tartary buckwheat were analyzed by Bonafaccia (2003). In both species most trace elements are concentrated mainly in the bran. However, there are relatively small differences in the contents of iron, antimony, and chromium between flour (extraction rate 55%) and bran fractions. In tartary buckwheat fine flour (extraction rate 42%) there is a lower trace element content than in normal flour. Thus the potential use of buckwheat bran as a dietary source of Zn, and Se, is postulated.

The hemicellulose fractions obtained by classical and ultrasound-assisted extraction exhibited significant immunomodulatory activities (Hromadkova and Ebringerova, 2003). It has been previously documented that the intake of buckwheat gives beneficial effects on hypertension and hypercholesterolemia (He *et al.*, 1995; Kayashita *et al.*, 1997). Buckwheat was found to have various bioactivities which include increasing lactic acid bacteria in rat intestine, treatment of allergic inflammation, reducing the serum glucose level, suppressing gallstone formation and cholesterol level, inhibiting the protease and scavenging radicals (Kayashita *et al.*, 1995; Kaw *et al.*, 2003; Kim *et al.*, 2003).

Secondary metabolites

Rutin is used medicinally in many countries to reduce capillary fragility associated with some hemorrhagic diseases or hypertension in humans (Yildzogle *et al.*, 1991).

Rutin is present at 2000 mg/100 g full weight in buckwheat sprouts and about 20 mg/100 g dry weight in the seeds, indicating that buckwheat sprouts synthesize a great amount of rutin during cotyledon growth (Kim *et al.*, 2004). On a dry weight basis, 10-day-old sprouts of tartary buckwheat, cultured with potting soil and wood shavings, contain 5–6% rutin, over twice that in common buckwheat sprouts (Kim *et al.*, 2006). Moreover, Watanabe and Shimizu (2004) have recently isolated and identified not only rutin but also C-glycosylflavones such as orientin, isoorientin, vitexin and isovitexin in 3-day-old tartary buckwheat sprouts (figure I). They also showed that while rutin in these sprouts increased with increasing periods of light exposure, C-glycosylflavone levels remained constant. Total phenol content and antioxidant activity was generally found to increase with sprouting, and a decrease in levels was observed following bread making (Alvarez-Jubete *et al.*, 2010). On average, about 300, 1000, and 46000 ppm of rutin were found in leaves, stems, and flowers, respectively. In case of seed, in bran fractions the concentration of rutin was 131-476 ppm, and in flour fractions 19-168 ppm (Kreft *et al.*, 1999).

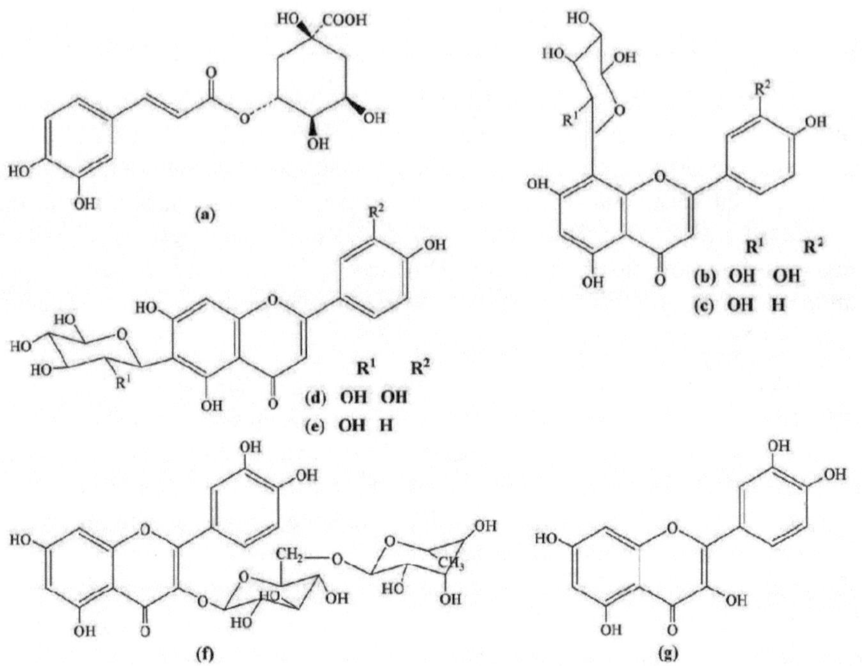

Fig. I: Structures of some secondary metabolites isolated from Buckwheat (a) chlorogenic acid (b) orientin (c) vitexin (d) isoorientin (e) isovitexin (f) rutin (g) quercetin

The variation of flavonoid content in sprouts of tartary buckwheat, grown in a greenhouse under low light conditions (16 µmol m^{-2} s^{-1}), was investigated by Kim *et al.* (2007). Chlorogenic acid and flavonoids including C-glycosylflavones (orientin, isoorientin, vitexin, isovitexin), rutin and quercetin were separated from the sprouts by HPLC and quantified with their commercial standards. Rutin content in the edible portion of the sprouts (mean 20 and 37 mg g^{-1} DW in 'Hokkai T 8' and 'Hokkai T 10', respectively) was 3- to 31-fold greater than that in the roots or pericarp. 'Hokkai T 8' and 'Hokkai T 10' are the two lines of tartary buckwheat, where 'Hokkai T 8' derived from the Russian 'Rotundatum' and, 'Hokkai T 10' obtained through ethyl methane sulfonate mutagenesis of 'Hokkai T 8'.

The physical and chemical properties of both common and tartary buckwheat seeds and products can greatly affected by different processing conditions. Physical and chemical qualities of buckwheat grit cakes were significantly influenced by processing conditions, including tempering moisture, heating temperature, and heating time (Im *et al.*, 2003). Thermal processing via autoclaving improved the total phenolic content, antioxidant activity and functionality of buckwheat (Randhir *et al.*, 2008). Roasting processing significantly reduced the protein quality and overall antioxidant properties of buckwheat whole seeds and groats (Zielinski *et al.*, 2009). Thermal treatment of tartary buckwheat flour caused a decrease in total phenolics, total flavonoid and antioxidative activities (Zhang *et al.*, 2010). The contents of polyphenols decreased during the preparation of tartary buckwheat bread due to the thermal treatment of tartary buckwheat flour. With the addition of water to mixtures containing tartary buckwheat during the preparation of the dough, rutin concentration decreased, while quercetin concentration increased. Rutin degradation could be caused by the rutin-degrading enzymes found in buckwheat (Vogrincic *et al.*, 2010).

Soak-treating tartary buckwheat significantly decreased the contents of total starch and rutin, whereas significantly increased the contents of quercetin, kaempferol, total flavonoid, and total phenolic compounds. In steam-treating soaked tartary buckwheat for 40–60 min, the contents of total starch, protein, crude fat, isoquercitrin, quercetin, kaempferol, total flavonoid and total phenolic compounds significantly decreased, whereas the content of rutin significantly increased. The reaction of rutin degrading might be catalyzed in soaking processing, and a catalytic reverse shift reaction of rutin-synthesizing might take place in steaming processing. The process of drying and roasting significantly decreased the contents of total starch, rutin, total flavonoid, and total phenolic compounds (Qin *et al.*, 2011). Tannin fractions were also obtained from the crude extracts of seeds and groats of buckwheat by Sephadex LH-20 column chromatography. Total phenolic contents of isolated fractions from buckwheat seeds and groats were 477 and 371 mg catechin equivalents/g, respectively (Karamac, 2010).

The flavonoid 3-O-glucosyltransferase was isolated and purified from buckwheat (*F. esculentum*) cotyledons. It was purified about 171-fold with final specific activities of 88.8 pmol of isoquercitrin (quercetin-3-glucoside) generation from quercetin per min per mg protein. The enzyme was determined to be a monomer by gel filtration (56.0 kDa) and electrophoresis (58.6 kDa). The 3-O-glucosyltransferase activity increased linearly together with rutin accumulation and then rapidly decreased when the rutin concentration reached a maximum value. The 3-O-glucosyltransferase activity and rutin concentration were high even when buckwheat cotyledons were grown under darkness (Suzuki *et al.*, 2005).

The rutin content of buckwheat products was compared to the rutin content in their raw materials, in order to evaluate their value for producing functional foods (Kreft *et al.*, 2006). There is much less rutin in noodles (78 mg/kg DW), than in the dark buckwheat flour (218 mg/kg DW) from which they are produced. One of the possible explanations is the presence of the rutin degrading enzyme. In raw (uncooked) groats there is 230 mg/kg DW of rutin and in precooked groats, 88 mg/kg DW. In buckwheat beer and vinegar there is a negligible content of rutin. Buckwheat leaf flour contains about 2700 mg/kg DW rutin. The rutin and flavonoid contents of both common and tartary buckwheat are compared and presented in table 4. The contents of rutin as well as total flavonoids were significantly different depending on species, 0.02% and 0.04% in *F. esculentum*, 0.10% and 0.35% in *F. homotropicum*, and 1.67% and 2.04% in *F. tataricum*, respectively. Three buckwheat species exhibited a dose response effect in inhibiting low-density lipoprotein (LDL) peroxidation. The antioxidant activity decreased in the order: *F. tataricum*> *F. homotropicum* >*F. esculentum*. It provides useful information for buckwheat breeding to develop high rutin content varieties (Jiang *et al.*, 2007). The LC–ESI-Q-TOFMS studies on whole buckwheat flour by Inglett *et al.* (2011) showed that p-coumaric and gallic acids were found in the bound phenolics along with isoquercitrin but were not present in the free phenolic compounds. The free flavonol- glycosides were found in whole buckwheat flour but not in any other buckwheat flours.

Whole buckwheat grains were milled into 16 flour fractions using the gradual milling system and the phenolic compounds and the antioxidant capacity of each flour fraction were investigated by Hung and Morita (2008). The phenolic and flavonoid contents of both free and bound phenolic extracts of buckwheat flour fractions significantly increased in the order from the fraction number 1 (phenolics less rich fraction) to the fraction number 16 (phenolics rich fraction). Table 5 shows the content of phenolics in different buckwheat materials. The phenolic compounds in buckwheat existed primarily in free form, whereas the flavonoids existed in grain in insoluble bound forms, bound to cell wall materials. As a result, the flour milled from the outer layers of buckwheat grains with large amount of phenolic compounds and antioxidant capacity are considered to have significant health benefits. In the edible parts of common buckwheat sprouts, individual phenolics significantly increased during sprout growth from 6 to 10 days after sowing (DAS), whereas in tartary buckwheat sprouts they did not. While the sum contents of phenolic compounds in the edible part (mean 24.4 mg/g DW at 6–10 DAS) of tartary buckwheat sprouts were similar to those of common buckwheat sprouts, rutin contents in the non-germinated/germinated seeds (mean 14.7 mg/g DW) and edible parts (mean 21.8 mg/g DW) of tartary buckwheat were 49- and 5-fold, respectively, higher than those of common buckwheat (Kim *et al.*, 2008). The optimal harvest time to enhance antioxidant and phytochemical content of buckwheat plants is found to be between 4–5 weeks after sowing.

Table 4: Rutin and flavonoid contents of common and tartary buckwheats (mg/kg DW)

	Rutin	Total flavonoids
Common buckwheat		
Dark flour[a]	218.5±7.1	—
Light flour[a]	112.8±7.7	—
Leaf flour[a]	2692.0±37.5	—
Groats precooked[a]	87.9±10.7	—
Groats tea[a]	204.3±18.4	- -
Raw groats[b]	297.6±10.0	- -
Roasted groats[b]	102.6±1.8	- -
Tartary buckwheat		
Raw whole seeds[c]	22593±60	22879±210
Soaked whole seeds[c]	10367±60	26933±230
Streamed whole seeds[c]	21643±130	21852±350
Dried whole seeds[c]	20652±110	21845±270
Dehulled groats[c]	11475±050	12991±160
Buckwheat tea[c]	11184±30	12090±290
Hull[d]	6342.1±9.2	8238.4±6.6
Groats[d]	9860.5±7.9	13869.7±8.2
Embryo[d]	24414.5±12.7	30091.9±19.0
Endospermum[d]	111.5±7.6	124.1±1.2
Testa[d]	61498.3±11.3	67994.7±14.9
Tartary buckwheat Sprouts		
Seed, 1 DAS[e]	12940±770	—
Seed, 2 DAS[e]	11980±1620	—
Edible part, 6 DAS[e]	16160±2100	2120±280
Edible part, 9 DAS[e]	29280±6120	4420±1510
Root, 6 DAS[e]	2990±620	—
Root, 9 DAS[e]	4170±770	—

[a]Kreft *et al.*, 2006; [b]B³aszczak *et al.*, 2011; [c]Qin *et al.*, 2011; [d]Li *et al.*, 2010; [e]Kim *et al.*, 2007

Recent studies have indicated that rutin intake improved antioxidant system in the rat liver, inhibited the increase in the level of thiobarbituric acid reactive substances in the gastric mucosa caused by ethanol injury, and suppressed oxygen radical overproduction in both rheumatoid arthritis and Fanconi anemia (Gao *et al.*, 2003; La-Casa *et al.*, 2000; Ostrakhovitch and Afanas'ev, 2001). The sequential separation and purification procedures for rutin from common buckwheat were also established by Kim *et al.* (2005) which involved extraction with 50% (v/v) aqueous ethanol at 80°C for 1 h followed by elution with water and aqueous ethanols at 20% and 30% (v/v) on a styrene-based resin column, and recrystallization at 4°C for 12 h. These conditions resulted in the recovery of 92% of total rutin with over 95% purity (Kim *et al.*, 2005).

The phytosterols were also found in buckwheat seeds and showed positive effects in lowering effect of blood cholesterol level, even at a low level. Plant sterols were reported to be able to perform antiviral effect so that can improve the immunological status of testing subjects (Li and Zhang, 2001). Fagopyrins were also found in the buckwheat seeds in very low concentrations which can be utilized in the treatment of type II diabetes (Krkoskova and Mrazova, 2005).

Five different isolation techniques were combined with gas chromatographic–mass spectrometric determination of aroma compounds from buckwheat: dynamic headspace with cryotrapping or sorbent trapping, solid-phase microextraction, headspace sorptive extraction, solvent extraction and simultaneous distillation extraction. Optimisation of each technique is presented, as well as comparison of the chemical profiles obtained by them. Solvent extraction with methanol resulted in the isolation of rather polar and less volatile compounds and was deemed least suitable for the odour active compounds. Only solid-phase microextraction was found to be suitable for the isolation of highly volatile compounds in a wider polarity range. The dynamic headspace extracted the smallest number of compounds, but with the comparable chemical profile as obtained by simultaneous distillation extraction, which was chosen as the most suitable technique to obtain the aroma compounds of cooked buckwheat (Prosen *et al.*, 2010). The compounds with the highest contribution to the buckwheat aroma were found to be of 2,5 – dimethyl - 4 – hydroxyl - 3(2H) - furanone, (E,E) - 2,4 - decadienal, phenylacetaldehyde, 2-methoxy-4-vinylphenol, (E)-2-nonenal, decanal, hexanal and salicylaldehyde (2-hydroxybenzaldehyde) (Janes *et al.*, 2009).

3. ANTINUTRITIONAL STUDIES

Tropane alkaloids have a long history of food-related toxicological issues intertwined with an established hallucinogenic background, as mentioned in most traditional pharmacopoeias (Christen, 2000; Huxtable, 1992). As a consequence of the high sensitivity of mammalians to these anticholinergic alkaloids, both in their racemic or enantiomeric pure forms, even extremely low amounts of atropine derivatives may trigger severe poisoning symptoms (Müller and Wanke, 1998). Few years ago an outbreak of food intoxication involving more than 60 people and related to *Fagopyrum esculentum* Moench. flour contaminated with scopolamine and atropine was reported in Slovenia and a subsequent food security alert was issued through the European Community (Perharic, 2005). Caligiani *et al.* (2011) detected toxic tropane alkaloids such as scopolamine and atropine in commercial buckwheat samples and related food products using gas-chromatography–mass-spectrometry in single-ion mode. The limits of detection for atropine and scopolamine were found to be 0.3 and 1 μg/kg, respectively. This contingency is of particular relevance for cultivations in which weed control is less strict and entrusted to laborious, manual operations like in organic agriculture (Perharic, 2005).

Eventhough there are many nutritional and medicinal health benefits from buckwheat and its products, some cases of allergic reactions were also reported. The main symptoms involved were asthma, allergic rhinitis, urticaria and angioeodema. Allergenic symptoms can occur even after exposition to buckwheat dust. For example, using pillows filled with buckwheat hulls, which were contaminated with buckwheat flour, can cause allergies (Li and Zhang, 2001; Bourrier *et al.*, 2003). In most cases of allergies the mechanism involved is an IgE related reaction (Li and Zhang, 2001). The major allergenic proteins in buckwheat have been revealed to be of molecular weight of 100, 50 and 17 kD (Li and Zhang, 2001; Urisu *et al.*, 1995). Immunoblotting using sera from a patient who suffered repeated episodes of anaphylaxis following buckwheat ingestion revealed four IgE-binding bands in the molecular weight range 9–40 kD, all of which were glycoproteins (Bush and Hefle, 1996). Aubrecht and Biacs (1999) carried out immunochemical analysis of buckwheat prolamins and investigated their allergenicity.

Buckwheat fibre is free of phytic acid (Steadman *et al.*, 2001). The trypsin inhibitors which are the common antinutritional factors found in most of the edible natural products, were isolated from buckwheat seeds extracts. These components are thermostable and can make digestion worse, if they are not adequately heat-treated before consumption. Besides trypsin these inhibitors can inhibit also chymotrypsin (Tsybina *et al.*, 2001).

4. ANTIOXIDANT STUDIES

In general, polar organic solvents are most effective for producing higher level of phenolic compounds and antioxidant activity (Przybylski *et al.*, 1998). For buckwheat, 80% methanol was found to extract 64 times more phenolic compounds and four times the antioxidant activity than water (Zielinski and Kozlowska, 2000). In another study, the effectiveness of solvents including methanol, ethanol, butanol, acetone, and ethyl acetate was investigated for phenolic compounds from buckwheat flour. Acetone extracted the most phenolic compounds, while methanol extracts had the highest antioxidant activity (Sun and Ho, 2005). In contrast, Gallardo *et al.* (2006) reported similar levels of phenolic compounds and higher antioxidant activity in aqueous extracts from buckwheat flour compared with 80% methanol extracts.

Sun and Ho (2005) evaluated the common buckwheat extracts for their antioxidant activities. The total phenolics were found to be higher in acetone extract which also showed highest scavenging activity of 78.6 ± 6.2% at 0.1 mg/ml by the DPPH method. The methanol extract showed the highest antioxidant activity coefficient of 627 ± 40.0 at 200 mg/l by the β-carotene bleaching method and longest induction time of 7.0 ± 0.2 h by the Rancimat method. Qin *et al.* (2011) performed the DPPH˙ scavenging activity of differently processed seeds of tartary buckwheat and found that soaked whole seeds possess higher scavenging activity than other types (table 5).

Table 5: The amount of total phenolics and DPPH˙ scavenging activities of common and tartary buckwheat (mg/g dry weight)

Plant material	Total Phenolics (mg/g DW)	DPPH˙ scavenging activity (µM Trolox Equivalents/g DW)
Common buckwheat[a]	81	
Groats raw[b]	—	18.29±1.21
Groats Roasted[b]	—	5.06±0.03
Sprout, 6 DAS[a]	—	1.5±0.2
Sprout, 9 DAS[a]	—	1.7±0.2
Tartary buckwheat[a]	90	
Raw whole seeds[c]	12.992±0.08	69.214±1.75
Soaked whole seeds[c]	16.224±0.34	92.987±3.15
Streamed whole seeds[c]	10.539±0.12	45.849±4.25
Dried whole seeds[c]	8.752±0.18	36.218±3.28
Dehulled groats[c]	5.412±0.23	21.896±2.72
Tartary Buckwheat tea[c]	4.751±0.25	19.286±2.92
Sprout, 6 DAS[d]	—	1.52±0.39
Sprout, 9 DAS[d]	—	2.14±0.02939

[a]Kim *et al.*, 2008; [b]B³aszczak *et al.*, 2011; [c]Qin *et al.*, 2011; [d]Kim *et al.*, 2007
DAS: Days After Sowing

Inglett in 2011 investigated commercial buckwheat flours for their antioxidant activities, free, and bound phenolic compositions. The free and bound phenolic contents were found to be in the decreasing order of Farinetta flour < Supreme < buckwheat < Fancy flour. The higher free phenolic content was achieved by 50% ethanol extraction compared with water and absolute alcohol whereas more bound phenolic and flavonoid compounds were found to be higher in absolute ethanol extract. It was also significant to note that a high extraction temperature increased the phenolic contents. The optimal solvent to solids ratio was standardized to 50:1, and the optimal extraction time was 5 h for flavonoids. Variation in antioxidant activity of buckwheat was mainly due to the cultivars and environment effects (Oomah and Mazza, 1996).

Antioxidant activities of buckwheat seeds, dehulled seeds, hulls, straws and leaves were evaluated by Holasovaa *et al.* (2002) and compared with that of oats and barley. Schaal oven test at 70°C with gravimetric indication and lard as the substrate was applied. Protection factors calculated to be ranged from 1.3 to 8.0 and increased in the order: buckwheat straws < buckwheat hulls = oats <barley <buckwheat seeds <buckwheat dehulled seeds <buckwheat leaves. Ethanol extracts from tartary buckwheat samples show that its DNA protective ability is higher than that from common buckwheat samples because of their high rutin, quercetin and phenolic content (Wei *et al.*, 2008). Ya-ping, (2008) studied the ethanol extract from the seeds of common buckwheat and tartary buckwheat for major antioxidant constituents. HPLC analysis of the extract showed two main antioxidation peaks, which correspondingly had same chromatographic retention times and spectral information to that of rutin and quercetin in the control samples.

Kim *et al.* (2007) studied the free radical scavenging activity in the edible portions of sprouts of tartary buckwheat. It was found that from 6 to 10 days after sowing, the DPPH radical scavenging rose significantly from 1.52 to 2.33 μM Trolox equivalents/g DW in 'Hokkai T 8' and from 1.46 to 2.09 μmol Trolox equivalents/ g DW in 'Hokkai T 10'.

Zhang *et al.* (2010) reported the effects of roasting, pressured-steam heating and microwave heating on total phenolics, total flavonoids, and antioxidant properties of whole-meal flour from tartary buckwheat (Table 6). It was also suggested that since the thermal treatment flour caused a decrease in total phenolics, total flavonoids and antioxidative activities of buckwheat, care should be taken when processing method is selected for the exploration of tartary buckwheat products.

Effect of processing on buckwheat flour was carried out before and after roasting or extrusion by Sensoy *et al.* (2006). Folin–Ciocalteu assay indicated that processing did not cause any change in total phenolic content in buckwheat flour. Roasted (200°C, 10 min) dark buckwheat flour exhibited an increase in non-polar compounds as well as polar compounds whereas extrusion exhibited increase only in polar compounds. Antioxidant activity test by DPPH showed that roasting at 200°C for 10 min decreased the antioxidant activity slightly whereas extrusion (170°C) did not cause any change. The ethanolic extract from tartary buckwheat sprout grown in 300 ppm Trace Element Water showed higher DPPH radical scavenging activity, ferrous ion chelating activity, superoxide anion scavenging activity and inhibitory activity toward lipid peroxidation than that grown in deionized water (Hsu *et al.*, 2008).

Buckwheat protein isolate was subjected to a two-stage *in vitro* digestion which ivnolved 1 h pepsin followed by 2-h pancreatin at 37°C. The 2-h pancreatin digest, which demonstrated the strongest activity against both radicals, was subjected to Sephadex G-25 gel filtration. Of the six fractions collected, fractions IV (456 Da) and VI (362 Da) showed the highest $ABTS^{+\bullet}$ scavenging activity. LC–MS/MS identified Trp-Pro-Leu, Val-Pro-Trp, and Val-Phe-Pro-Trp (IV), Pro-Trp (V) and tryptophan (VI) to be the prominent peptides/ amino acid in these fractions (Maa *et al.*, 2010).

Table 6: Effect of processing on Hydroxyl radical and Superoxide radical scavenging activities of tartary buckwheat flour

Different processing	Hydroxyl radical scavenging activity %	Superoxide radical scavenging activity %
Raw flour	93.13±2.58	92.64±2.22
80°C, 20 min	87.27±2.80	82.67±2.45
80°C, 40 min	87.90±3.75	77.70±3.22
120°C, 20 min	76.33±4.80	72.40±4.40
120°C, 40 min	68.53±9.07	63.83±6.96
0.1MPa, 20 min	29.83±2.63	74.00±7.46
0.1MPa, 40 min	20.17±5.75	63.33±2.91
0.2MPa, 20 min	12.26±1.67	54.13±3.77
0.2MPa, 40 min	12.10±1.67	47.37±0.46
microwave, 10 min	30.63±0.91	71.47±0.65

5. PHARMACOLOGICAL STUDIES

Rufa (2001) reported that the extract of tartary buckwheat had no negative effect on the development and the indexes of hematology, biochemistry and pathology after continuous administration for a long time. Kim *et al.* (2003) investigated the anti-allergic action of buckwheat grain extract using rodent experimental models. The oral, intraperitoneal and intradermal administration of extract significantly inhibited the compound 48/80-induced vascular permeability documented by Evans blue extravasation. In addition, buckwheat grain extract showed potent inhibitory effect on passive cutaneous anaphylaxis (PCA) activated by anti-dinitrophenyl (DNP) IgE when orally administered. In an *in vitro* study, grain extract revealed to possess inhibitory potential on the compound 48/80-induced histamine release from rat peritoneal mast cells (RPMC). Moreover, grain extract inhibited the IL-4 and TNF-a mRNA induction by PMA and A23187 in human leukemia mast cells, HMC-1. Taken together, these results suggest that anti-allergic action of grain extract may be due to the inhibition of histamine release and cytokine gene expression in the mast cells.

Han and Zhang (2004) reported that total flavones content of buckwheat flower has some protective effects on isoproterenol-induced cardiac hypertrophy in rats. A study was conducted to investigate the hypolipidemic effects of a powdered whole buckwheat leaf and flower mixture in rats fed a high-fat diet by Lee *et al.* (2010). The result suggested that the beneficial effect of this buckwheat plant portion on plasma and hepatic lipid profiles in high-fat fed rats is partly mediated by higher excretion of faecal lipids and synergistic effect of phenolic compounds and fibre present in the buckwheat.

There are several experimental studies, comparing health effects in animals fed with either buckwheat protein extract diet, casein, or soy protein diet. Lower plasma cholesterol and less gall stone formation in hamsters was observed with protein extract diet, as compared to casein or soya protein diet (Tomotake *et al.*, 2000). The mechanism was suggested to be an enhancement of bile acid synthesis and faecal excretion of steroids. The buckwheat protein extract caused muscle hypertrophy and reduced body fat content, as compared to rats fed on casein (Kayashita *et al.*, 1999). It has also been shown that protein extract ameliorates atropine-induced constipation in rats (Kayashita *et al.*, 1995). Kayashita *et al.* (1997) made another animal study, suggesting that the cholesterol lowering effect of protein extract is mediated by higher faecal excretion of neutral sterols and that lower digestibility of buckwheat protein is partly responsible for the effect. Buckwheat herb shows significant anti oedema effects on artificially produced leg oedema in female guinea pigs (Wanderka, 1981). In addition, inhibitory effect on lipid peroxidation was observed in rabbits fed with a fatty diet (Wojciki, 1995). Mice fed on extracts of tartary buckwheat leaf had increased enzyme activities of superoxide dismutase (SOD) and GSH-peroxidase in blood, liver and the heart, and increased activity of catalase in many organs, suggesting that tartary buckwheat leaf extract has a high potential in antioxidation and anti-lipid peroxidation (Wang *et al.*, 1998). Finally, it was shown that buckwheat protein diet may retard the development of mammary tumours in rats, exposed to an experimental carcinogen. Less mammary tumours and lower serum estradiol was found after feeding with buckwheat protein extract, as compared to casein diet. The anti-carcinogenic effect was proposed to be mediated by the lowering of serum estradiol levels at buckwheat protein diet (Kayashita *et al.*, 1999).

6. CONCLUSIONS

Buckwheat has been praised as one of the most faddish green functional foods of the 21st century (Piao and Li, 2001; Zhang and Chen, 2004). The nutritional and antioxidant properties alone can bring both common and tartary buckwheat to the position where functional food meets the demand. The anti-nutritional and allergic factors of buckwheat have not yet been elicited entirely. Hence a step is needed to sort out these factors which will designate the plant more close to the safety food. Even though buckwheat forms the major source of rutin, the development of simpler, easier and economical isolation methods can only highlight buckwheat to become the commercial source of rutin. The presence of higher amount of flavonoids, phenolics and bioactive compounds make buckwheat as one of the good candidate for extensive pharmacological research which is not much explored till date. *In vivo* studies like acute toxicity, sub acute toxicity, chronic toxicity etc has to be carried out with the edible parts or with biologically active plant extracts for the safety evaluation of buckwheat. Even though there are lot of breeding experiments and micropropagation programmes done in buckwheat to promote its cultivation, there is a big gap between buckwheat and common people or crop cultivators. One major problem with buckwheat today is that, it has become an issue of food status which is being restricted to the diet of poor community only, so scientific validation and awareness about the nutritious and medicinal status of buckwheat can only bringout its wide utilization. Therefore, effective research should be done to explore and improve the indigenous characters of buckwheat which seemed to provide potential nutraceuticals for human health. For countries with wide population and also suffering from food scarcity, time has come to focus on such under- utilized neglected nutritious crops to meet the demand for food security.

REFERENCES

Alvarez-Jubete, L., Wijngaard, H., Arendt, K., Gallagher, E., 2010. Polyphenol composition and *in vitro* antioxidant activity of amaranth, quinoa buckwheat and wheat as affected by sprouting and baking. *Food Chem.*, 119:770–778.

Aubrecht, E., Biacs, P.A., 1999. Immunochemical analysis of buckwheat proteins, prolamins and their allergenic character. *Acta Aliment.*, 28(3):261–268.

B³aszczak, W., Zieliñska, D., Zieliñski, H., Szawara-Nowak, D., Fornal, J., 2011. Antioxidant Properties and Rutin Content of High Pressure-Treated Raw and Roasted Buckwheat Groats. *Food Bioprocess Technol* DOI 10.1007/s11947-011-0669-5.

Bonafaccia, G., Fabjan, N., 2003. Nutritional comparison of tartary buckwheat with common buckwheat and minor cereals. *Zb Bioteh Fak Univ Ljublj Kmet.*, 81: 349-355.

Bonafaccia, G., Marocchini, M., Kreft, I., 2003. Composition and technological properties of the flour and bran common and tartary buckwheat. *Food Chem.,* 80:9–15.

Bonafacciaa, G., Gambellia, L., Fabjanb, N., Kreft, I., 2003. Trace elements in flour and bran from common and tartary buckwheat. *Food Chem.*, 83:1–5.

Bourrier, T., Noormahomed, T., Coache, M., Albertini, M., Boutte, P., 2003. A case of buckwheat allergy in a child. *Revue Française d'allergologie et d'immunologie Clinique.*, 43:530–532.

Bush, R.K., Hefle, S.L., 1996. Food allergens. *Crit Rev Food Sci Nutr.*, 36(S):S69–S89.

Caligiani, A., Palla, G., Bonzanini, F., Bianchi, A., Bruni, R., 2011. A validated GC–MS method for the detection of tropane alkaloids in buckwheat (*Fagopyron esculentum* L.) fruits, flours and commercial foods. *Food Chem.*, 127:204–209.

Campbell, C.G., 1997. Buckwheat. *Fagopyrum esculentum* Moench. Promoting the conservation and use of underutilized and neglected crops. 19. Institute of Plant Genetics and Crop Plant Research, Gatersleben/International Plant Genetic Resources Institute, Rome, Italy.

Christen, P., 2000. Tropane alkaloids: Old drugs used in modern medicine. In: Attaur- Rahman (ed). Studies in Natural Product Chemistry. Amsterdam: Elsevier, 22: pp.717–750.

Fujino, K., Funatsuki, H., Inada, M., Shimono, Y., Kikuta, Y., 2001. Expression, cloning, and immunological analysis of buckwheat (*Fagopyrum esculentum* Moench) seed storage proteins. *J Agric Food Chem.*, 49:1825–1829.

Gallardo, C., Jimenez, L., Garcia-Conesa, M.T., 2006. Hydroxycinnamic acid composition and in vitro antioxidant activity of selected grain fractions. *Food Chem.*, 99:455-463.

Gao, Z., Xu, H., Chen, X., Chen, H., 2003. Antioxidant status and mineral contents in tissues of rutin and baicalin fed rats. *Life Sci.*, 73:1599–1607.

Guo, X.N., Yao, H.Y., 2005. The studies of structure, low digestibility and physiological functions on buckwheat protein. *Machinery cereal oil Food Process*, pp 60–64.

Guo, X.N., Yao, H.Y., 2006. Fractionation and characterization of tartary buckwheat flour proteins. *Food Chem.*, 98:90–94.

Guo, Y.Z., Chen, Q.F., Yang, L.Y., Huang, Y.H., 2007. Analyses of the seed protein contents on the cultivated and wild buckwheat *Fagopyrum esculentum* resources. *Genet Resour Crop Evol.*, 54:1465–1472.

Han, S., Zhang, J., 2004. Protective effects of total flavones of buckwheat flower on experimental cardiac hypertrophy in rats. *J Forth Mili Med Uni.*, 25(14):1338–1340.

He, J., Klag, M.J., Whelton, P.K., Mo, J.P., Chen, J.Y., *et al.*, 1995. Oats and buckwheat intakes and cardiovascular disease risk factors in an ethnic minority of China. *Am J Clin Nutr.*, 61:366–72.

Holasovaa, M., Fiedlerovaa, V., Smrcinovaa, H., Orsakb, M., Lachmanb, J., Vavreinovaa, S., 2002. Buckwheat-the source of antioxidant activity in functional foods. *Food Res Int.*, 35:207–211.

Hromadkova, Z., Ebringerova, A., 2003. Ultrasonic extraction of plant materials—investigation of hemicellulose release from buckwheat hulls. *Ultrason Sonochem.*, 10:127–133.

Hsu, C.K., Chiang, B.H., Chen, Y.S., Yang, J.H., Liu, C.L., 2008. Improving the antioxidant activity of buckwheat (*Fagopyrum tataricm* Gaertn) sprout with trace element water. *Food Chem.*, 108:633–641.

Hung, P.V., Morita, N., 2008. Distribution of phenolic compounds in the graded flours milled from whole buckwheat grains and their antioxidant capacities. *Food Chem.*, 109:325–331.

Huxtable, R.J., 1992. The toxicology of alkaloids in foods and herbs. In: Tu AA (ed.). Food Poisoning, Handbook of Natural Toxins. New York: Marcel Dekker Inc, 7: pp. 252–253.

Im, J.S., Huff, H.E., Hsieh, F.H., 2003. Effects of processing conditions on the physical and chemical properties of buckwheat grit cakes. *J Agric Food Chem.*, 51:659"666.

Inglett, G.E., Chen, D., Berhow, M., Lee, S., 2011. Antioxidant activity of commercial buckwheat flours and their free and bound phenolic compositions. *Food Chem.*, 125:923–929.

Janes, D., Kantar, D., Kreft, S., Prosen, 2009. Identiûcation of buckwheat (*Fagopyrum esculentum* Moench) aroma compounds with GC–MS. *Food Chem.*, 112:120–124.

Jiang, P., Burczynski, F., Campbell, C., Pierce, G., Austria, J.A., Briggs, C.J., 2007. Rutin and xavonoid contents in three buckwheat species *Fagopyrum esculentum*, *F. tataricum*, and *F. homotropicum* and their protective effects against lipid peroxidation. *Food Res Int.*, 40:356–364.

Joshi, B.D., Paroda, R.S., 1991. Buckwheat in India. NBPGR, Shimla. Sci Monogr No.2: pl17.

Karamac, M., 2010. Antioxidant Activity of Tannin Fractions Isolated from Buckwheat eeds and Groats. *J Am Oil Chem Soc.*, 87:559-566.

Kawa, J.M., Taylor, C.G., Przybylski, R., 2003. Buckwheat concentrate reduces serum glucose in streptozotocin-diabetic rats. *J Agric Food Chem.*, 51:7287–7291.

Kayashita, J., Shimaoka, I., Nakajoh, M., 1995. Hypocholesterolemic effect of buckwheat protein extract in rats fed cholesterol enriched diets. *Nutr Res.*, 15:691–698.

Kayashita, J., Shimaoka, I., Nakajoh, M., Kishida, N., Kato, N., 1999. Consumption of a buckwheat protein extract retards 7,12-dimethylbenz [a] anthracene-induced mammary carcinogenesis in rats. *Biosci Biotechnol Biochem.*, 63:1837–1839.

Kayashita, J., Shimaoka, I., Nakajoh, M., Kondoh, M., Hayashi, K., Kato, N., 1999. Muscle hypertrophy in rats fed on Buckwheat protein Extract. *Biosci Biotechn Biochem.*, 63:1242-1245.

Kayashita, J., Shimaoka, I., Nakajoh, M., Yamazaki, M., Kato, N., 1997. Consumption of buckwheat protein lowers plasma cholesterol and raises fecal neutral sterols in cholesterol-fed rats because of its low digestibility. *J Nut.*, 127:1395–400.

Kayashita, J., Shimaoka, I., Nakajoh, M., Yamazaki, M., Kato, N., 1997. Consumption of buckwheat protein lowers plasma cholesterol and raises faecal neutral sterols in cholesterol- fed rats because of its low digestibility. *J Nut.*, 127:1395-1400.

Kayashita, J., Shimaoka, I., Yamazaki, M., Kato, N., 1995. Buckwheat protein extract ameliorates atropine - induced constipation in rats. In: Matano T, Ujihara A (eds.) Current Advances in Buckwheat Research. Vol II, Shinshu University Press, Matsomoto, pp. 941-946.

Kim, C.D., Leec, W.K., Noc, K.O., Parkc, S.K., Leec, M.H., Lima, S.R., Roha, S.S., 2003. Anti-allergic action of buckwheat (*Fagopyrum esculentum* Moench) grain extract. *Intal Immunopharmacol.*, 3:129-136.

Kim, K.H., Lee, K.W., Kim, D.Y., Park, H.H., Kwon, I.B., Lee, H.J., 2005. Optimal recovery of high-purity rutin crystals from the whole plant of *Fagopyrum esculentum* Moench (buckwheat) by extraction, fractionation, and recrystallization. *Bioresource Technol.*, 96:1709-1712

Kim, S.J., Kawaharada, C., Suzuki, T., Saito, K., Hashimoto, N., Takigawa, S., Noda, T., Matsuura-Endo, C., Yamauchi, H., 2006. Effect of natural light periods on rutin, free amino acid and vitamin C contents in the sprouts of common (*Fagopyrum esculentum* Moench) and tartary (*F. tataricum* Gaertn.) buckwheats. *Food Sci Technol Res.*, 12:199–205.

Kim, S.J., Zaidul, I.S.M., Maeda, T., Suzuki, T., Hashimoto, N., Takigawa, S., Noda, T., Matsuura-Endo, C., Yamauchi, H., 2007. A time-course study of flavonoids in the sprouts of tartary (*Fagopyrum tataricum* Gaertn.) buckwheats. *Sci Hort.*, 115:13–18.

Kim, S.J., Zaidul, I.S.M., Suzuki, T., Mukasa, Y., Hashimoto, .N, Takigawa, S., Noda, T., Matsuura-Endo, C., Yamauchi, H., 2008. Comparison of phenolic compositions between common and tartary buckwheat (*Fagopyrum*) sprouts. *Food Chem.*, 110:814–820.

Kim, S.L., Kim, S.K., Park, C.H., 2004. Introduction and nutritional evaluation of buckwheat sprouts as a new vegetable. *Food Res Int.*, 37: 319–327.

Kreft, I., Fabjan, N., Yasumoto, K., 2006. Rutin content in buckwheat (*Fagopyrum esculentum* Moench) food materials and products. *Food Chem.*, 98:508–512.

Kreft, S., Knapp, M., Kreft, I., 1999. Extraction of Rutin from Buckwheat (*Fagopyrum esculentum* Moench) Seeds and Determination by Capillary Electrophoresis. *J Agric Food Chem.*, 47:4649"4652.

Krkoskova, B., Mrazova, Z., 2005. Prophylactic components of buckwheat. *Food Res Int.*, 38:561–568.

La-Casa, C., Villegas, I., Alarcon-de-la-Lastra, C., Motilva, V., Martin-Calero, M.J., 2000. Evidence for protective and antioxidant properties of rutin, a natural Xavone, against ethanol induced gastric lesions. *J Ethnopharmacol.*, 71:45–53.

Ladefoged, A.V.N., 1952. Cultivation of bucwkheat in up country Patne. *Trop Agric.*, 10:269-270.

Lee, J.S., Bok, S.H., Jeon, S.M., Kim, H.J., Do, K.M., Park, Y.B., Choi, M.S., 2010. Antihyperlipidemic effects of buckwheat leaf and flower in rats fed a high-fat diet. *Food Chem.*, 119:235-240.

Li, D., Li, X., Ding, X., 2010. Composition and antioxidant properties of the flavonoid-rich fractions from tartary buckwheat grains. *Food Sci Biotechnol.*, 19(3):711-716.

Li, S., Zhang, Q.H., 2001. Advances in the development of functional foods from buckwheat. *Food Sci Nutr.*, 41(6):451-464

Maa, Y., Xiong, Y.L., Zhai, J., Zhu, H., Dziubla, T., 2010. Fractionation and evaluation of radical scavenging peptides from *in vitro* digests of buckwheat protein. *Food Chem.*, 118:582-588.

Maeda, T., Miyake, K., Tahara, M., Morita, N., 2004. Substitution of buckwheat for wheat flours on processing pasta and cookie. *Fagopyrum,* 21:99-103.

McGregor, W.G., McKillican, M.E., 1952. Rutin content of varieties of buckwheat. *Sci Agri.*, 32:48-51.

Milisavljevic, M.D., Timotijevic, G.S., Radovic, S.R., Brkljacic, J.M., Konstantinovic, M.M., Maksimovic, V.R., 2004. Vicilin-like storage globulin from buckwheat (*Fagopyrum esculentum* Moench) seeds. *J Agric Food Chem.*, 52:5258-5262.

Müller, J., Wanke, K., 1998. Toxic psychoses from atropine and scopolamine. *Fortschr Neurol Psychiatr.*, 66:289-295.

Oomah, B.D., Mazza, G., 1996. Flavonoids and antioxidative activities in buckwheat. *J Agric Food Chem.*, 44:1746-1750.

Ostrakhovitch, E.A., Afanas'ev, I.B., 2001. Oxidative stress in rheumatoid arthritis leukocytes: Suppression by rutin and other antioxidants and chelators. *Biochem Pharmacol.*, 62:742–746.

Park, C.H., Kim, Y.B., Choi, Y.S., Heo, K., Kim, S.L., Lee, K.C. *et al.*, 2000. Rutin content in food products processed from groats, leaves and flowers of buckwheat. *Fagopyrum,* 17:63-66.

Perharic, L., 2005. Mass tropane alkaloid poisoning due to buckwheat flour contamination. *Clin Toxicol.*, 43:413.

Piao, S.L., Li, L.H., (eds), 2001. The actuality of produce and exploitation of *Fagopyrum* in China. Proceeding of the 8th International Symposium on buckwheat, Chunchon, Korea. *Adv Buckwheat Res.*, 571-576.

Pomeranz, Y., Robbins, G.S., 1972. Amino acid composition of buckwheat. *J Agric Food Chem.*, 20:270-274.

Prosen, H., Kokalj, M., Janeš, D., Kreft, S., 2010. Comparison of isolation methods for the determination of buckwheat volatile compounds. *Food Chem.*, 121:298-306.

Przybylski, R., Lee, Y.C., Eskin, N.A.M., 1998. Antioxidant and radical-scavenging activities of buckwheat seed components. *J Am Oil Chem Soc.*, 75:1595-1601.

Qin, P., Wu, L., Yao, Y., Ren, G., 2011. Changes in phytochemical compositions, antioxidant and á- glucosidase inhibitory activities during the processing of tartary buckwheat tea. *Food Res Int* doi:10.1016/j.foodres.2011.03.028.

Randhir, R., Kwon, Y., Shetty, K., 2008. Effect of thermal processing on phenolics, antioxidant activity and health-relevant functionality of select grain sprouts and seedlings. *Innovat Food Sci Emerg Tech.*, 9:355"364.

Rufa, L., Yunning, Z., Rui, W., Jianying, L., 2001. A Study on the Extract of Tartary Buckwheat, The proceeding of the 8'hISB:602-607.

Sensoy, I., Rosen, R.T., Ho, C.T., Karwe, M.V., 2006. Effect of processing on buckwheat phenolics and antioxidant activity. *Food Chem.*, 99:388-393.

Steadman, K.J., Burgoon, M.S., Lewis, B.A., Edwardson, S.E., 2001. Minerals, phytic acid, tannin and rutin in buckwheat seed milling fractions. *J Sci Food Agri.*, 81:1094-1100.

Sun, T., Ho, C.T., 2005. Antioxidant activities of buckwheat extracts. *Food Chem.*, 90:743-749.

Suzuki, T., Kim, S.J., Yamauchi, H., Takigawa, S., Honda, Y., Mukasa, Y., 2005. Characterization of a flavonoid 3-O-glucosyltransferase and its activity during cotyledon growth in buckwheat (*Fagopyrum esculentum*). *Plant Sci.*, 169:943-948.

Tomotake, H., Shimaoka, I., Kayashita, J., Yokoyama, F., Nakajoh, M., Kato, N., 2000. A Buckwheat protein product suppresses gallstone formation and plasma cholesterol more strongly than soy protein isolate in hamsters. *J Nutr.*, 130:1670-1674.

Tsuji, K., Ohnishi, O., 2001. Phylogenetic relationships among wild and cultivated tartary buckwheat (*Fagopyrum tataricum* Gaert.) populations revealed by AFLP analyses. *Genes Genet Syst.*, 76:47-52.

Tsybina, T.A., Dunaevsky, Y.E., Musolyamov, A.K., Egorov, T.A., Belozersky, M.A., 2001. Cationic inhibitors of serine proteinases from buckwheat seeds. *Biochem.*, 66(9):943-947.

Urisu, A., Kondo, Y., Morita, Y., Yagi, E., Tsuruta, M., Yasaki, T. *et al.*, 1995. Isolation and characterization of a major allergen in buckwheat seeds. In: Matano T, Ujihara A (eds.) Current Advances in Buckwheat Research. Shinshu University Press, Matsomoto, pp. 965–974.

Vogrincic, M., Timoracka, M., Melichacova, S., Vollmannova, A., Kreft, I., 2010. Degradation of rutin and polyphenols during the preparation of tartary buckwheat bread. *J Agric Food Chem.*, 58:4883"4887.

Wanderka, H., 1981. Prüufung des phytotherpeuticums "Fagorutin" am küunstlich erzeugten staungodem. *Zeitschrift für Phytotherapie*, 2:225-233 (in German).

Wang, Z.H., Zang, Z., Lin, R.F., 1998. A study of the biological effects on the extracts of tartary buckwheat leaf. Section III. Utilization. The Seventh International Symposium on Buckwheat, Winnipeg, Canada, pp. 117-121.

Watanabe, M., 1998. Catechins as antioxidants from buckwheat (*Fagopyrum esculentum* Moench.) groats. *J Agric Food Chem.*, 46(3):839-845.

Watanabe, M., Shimizu, H., 2004. Composition of flavonoid compounds in seedlings of tartary buckwheat. *Tohoku Agric Res.*, 57:267-268.

Wei, C., Wei-Jun, C., Zhi-Rong, S., Ya-Ping, Y., 2008. Protective effects of ethanolic extracts of buckwheat groats on DNA damage caused by hydroxyl radicals. *Food Res Int.*, 41:924–929.

Wojciki, J., Barcew-Wiszniewska, B., Samochowiec, L., Rozewicka, L., 1995. Extractum fagopyr I reduces athreosclerosis in high-fat diet fed rabbits. *Die Pharmazie.*, 50:560-562.

Ya-ping, Y., Cheng-rui, T., Wei, C., 2008. Anti-Oxidative Constituents of Ethanol Extract from Buckwheat Seeds by HPLC- Electron spray MS. *Agric Sci China.*, 7(3):356-362.

Yildzogle, A.N., Altan, V.M., Altinkurt, O., Ozturk, Y., 1991. Pharmacological effects of rutin. *Phytother Res.*, 5:19-23.

Zhang, M., Chen, H., Li, J., Pei, Y., Liang, Y., 2010. Antioxidant properties of tartary buckwheat extracts as affected by different thermal processing methods. *Food Sci Technol.*, 43:181–185.

Zhang, Y.Z., Chen, Q.F., 2004. The actuality and expectation on buckwheat studies. *Seeds*, 23:39–42.

Zielinski, H., Kozlowska, H., 2000. Antioxidant activity and total phenolics in selected cereal grains and their different morphological fractions. *Journal of Agricultural and Food Chemistry*, 48:2008–2016.

Zielinski, H., Michalska, A., Amigo-Benavent, M., Castillo, M.D., Piskula, M.K., 2009. Changes in protein quality and antioxidant properties of buckwheat seeds and groats induced by roasting. *J Agric Food Chem.*, 57:4771-4776.

23

Quality Control by Avoiding Excess Nutrient Leaching

G. Vanitha, Mrinalini J. Singh and S. Padmavathy

1.INTRODUCTION

The total sum of foods or mixture of foods which an individual consumes each day is referred to as his diet (Okey, 1992). Vegetables that are used as food has nourishing substances called nutrients namely carbohydrates, proteins, fats, minerals and vitamins. The nutritive value of different vegetables varies sufficiently and the usage of variety of vegetables ensures all the necessary nutrients required in the diet and its lack in daily menu may result in various disease and deficiencies (Saxena, 1970). They contain a great variety of all other phytochemicals, some of which claimed to have antioxidant, antimicrobial and also anti-carcinogenic properties (Grudha et al., 1996). But when soaking is done in water either plain or salted with sodium chloride, water soluble nutrients may be leached into the water (Srilakshmi, 1999). An attempt was made in the present study to find out the effect of soaking of some selected vegetables in water for one hour after cutting them into pieces and its nutrient loss.

2. MATERIALS AND METHODS

Vegetables used for the present study were bitter gourd (*Momordiacharantia* L.), plantain (*Musa paradisiaca* L.), bean (*Phaseolus vulgaris* L.), brinjal (*Solanummelongena* L.), and drum stick (*Moringaolefera* Lam.).

Vegetables mentioned above were purchased from the market fresh on the day of analysis. They were estimated for some biochemical components. For each vegetable, a known amount was taken, cut into small pieces and kept immersed in known quantity of distilled water for one hour. Then, those cut vegetables were removed from the soaked water and were reanalyzed for the same biochemical components as in the fresh vegetables. In addition, the water in which the vegetables were soaked was also analyzed for the presence of some biochemical compounds in the leachate.

Estimation of starch was done by Iodine method (Meyer *et al.*, 1966) where 1 gm of vegetable tissue was thoroughly ground with 20 ml of distilled water and boiled over a waterbath for 15 minutes. The boiled homogenate was centrifuged at 2000rpm for 5 minutes and the clear supernatant was saved. To 1ml of sample, 8.5ml of distilled water and 0.5 ml of iodine reagent were added and mixed thoroughly and the color developed was measured in a calorimeter at 630 nm.

Estimation of reducing sugar was done (Nelson - Somogyi method, 1944) by boiling 1 ml of vegetable extract with 1 ml of distilled water and 1 ml of alkaline copper tartrate reagent until the development of orange color. After adding 1 ml of arsenomolybdate reagent,the color developed was read at 620 nm using calorimeter.

Estimation of protein was done (Lowry *et al.*, 1951) wherein 100 mg of sample was homogenized with 5 ml of ice cold phosphate buffer and centrifuged at 2000 rpm for 5 minutes. To the supernatant solution, equal volume of 10% ice cold TCA was added and incubated for 10 minutes at 4°C. The precipitated protein was centrifuged and the pellet was dissolved in 1 ml of 0.1N NaOH. To 0.5 ml of the protein solution, 5ml of alkaline copper reagent was added. It was shaken well and allowed to stand at room temperature for 10 minutes. Then, 0.5 ml of folin – ciocalteau reagent was added and the volume was made upto a known quantity using distilled water. Blank was prepared without the sample extract. After 30 minutes, the optical density of the solution was read at 660 nm in Spectronic – 20D.

Estimation of total free amino acid was done by Ninhydrin method (Moore & Stein, 1948) where 1 ml of alcoholic vegetable extract was boiled with 0.5 ml of acetate buffer and 1 ml of 1% Nihydrinfor color development which was measured at 570 nm calorimetrically.

Estimation of ascorbic acid was done by iodine method (Mahadevan and Sridhar, 1982) where 5ml of vegetable extract was titrated against 0.1% DCPIP dye till the appearance of pink color which persists for a few minutes. Standard ascorbate was also titrated against the dye. Using the following formula, the amount of ascorbate can be found.

$$\text{Ascorbate in mg/100g of the sample} = \frac{0.5 \text{ mg x} V2 \text{ x } 10\text{ml x } 100\text{g}}{V1 \text{ x } 5\text{ml x sample weight}}$$

V1 – Titre value of ascorbic acid

V2 – Titre value of the sample

10ml- made up quantity of the extract

Estimation of amylase activity was done by (Malik and Singh, 1980). To 1 ml of 0.1% starch solution, 1 ml supernatant of the centrifuged vegetable extract was added. After incubating for 5 minutes at 37°C, the reaction was arrested by adding 1 drop of conc. HCL. Then 0.5 ml of iodine was added to the reaction mixture and was made upto 10 ml with distilled water. The intensity of color was read at 620 nm in the spectronic -20.

Estimation of total phenol was done by Folin-Ciocalteau method (Mahadevan and Sridhar, 1982) where 1 ml supernatant of the centrifuged vegetable extract was heated with 1 ml of folinciocalteau reagent and 2 ml of 20% sodium carbonate. The absorbency was read at 650 nm using spectrophotometer.

3. RESULT AND DISCUSSION

In the present investigation, an experiment was conducted to find out the effect of soaking of vegetables in water for 2 hours and to analyze some nutrient composition of fresh and soaked vegetables. In addition, the loss of nutrients was analyzed in the water in which the cut vegetables were soaked.

The carbohydrates analyzed were total starch and reducing sugars. As the analyzed materials were vegetables, they contained little amount of starch. The reducing sugar content was found several folds greater than the starch. However, both of their content showed a decline after soaking.

Vegetables were estimated for their protein and amino acid content. The protein content of all the vegetables under study showed higher amount as compared to their respective starch content. The total free amino acid content of all the vegetables was recorded to be approximately 3 to 7 folds higher than their protein content. Obviously, the amino acid content of all the cut vegetables was found greater than the fresh vegetables, which could be attributed to the hydrolysis of protein into amino acid.

The secondary metabolite phenol also was analyzed. A considerable quantity was estimated in all the selected vegetables, however, it was noticeably high in brinjal which leads to its coloration after cutting. A substantial level of reducing sugar, amino acid and phenol were obtained in all leachates. Phenol was also estimated in remarkable quantity.

Amylase activity of the vegetable leachates was significantly high in bitter gourd, plantain followed by drumstick compared to other plants.

Ascorbic acid which possesses rich antioxidant activity was relatively higher value in bitter guard (59 mg/100g). In other vegetables, it ranged between 3 mg/100g to 6mg/100g. But its level exhibited a remarkable decline after soaking in water.

Parameters		Selected Vegetables				
		Bitter gourd	Plantain	Bean	Brinjal	Drumstick
Starch(µg/g)	Fresh	60	120	78	48	58
	Soaked	48	52	50	40	28
Reducing sugar(µg/g)	Fresh	1050	1400	3500	4590	4410
	Soaked	550	800	3400	1450	1900
	Leachates of cut veg	300	325	1300	825	900
Protein(µg/g)	Fresh	55	140	325	138	155
	Soaked	48	75	175	115	125
Amino acid(µg/g)	Fresh	600	900	950	1000	1500
	Soaked	1000	2000	2000	1075	1250
	Leachates of cut veg	400	800	900	550	1000
Phenol(µg/g)	Fresh	375	300	400	800	325
	Soaked	350	200	338	530	200
	Leachates of cut veg	50	250	55	100	150
Amylase (mg/h/g)	Leachates	28	26.7	7.5	4.25	18
Ascorbate (mg/100g)	Fresh	59	3.33	4.45	4.0	6.1
	Soaked	27	2.16	1.57	1.86	3.93

4. CONCLUSION

From the above observations, it is noted that cut vegetables if soaked for a long duration release lot of nutrients through the cut ends and at the same time, the hydration of tissues enhance the activity of the various enzymes. Hydrolysis would result in the breakdown of stored organic compounds like starch, protein and fat and lead to leaching of all nutrients and ultimately the nutrient less food would be taken for consumption.

Hence, as stated by number of dieticians, the vegetables that can be eaten raw need not be subjected to cutting and soaking for some duration. Vegetables that need cooking could be washed before cutting and after cutting, they can be soaked in small quantity of water and the same water could be used for cooking. Such kind of processing of vegetable might avoid excessive loss of nutrients from vegetables.

REFERENCES

Bernard S. Meyer., Donald B. Anderson and Carroll A.Swanson.1966. Laboratory Plant Physiology. Third Edition. Affiliated East-West Press, Pvt. Ltd., New Delhi. pp. 164-166.

Grudha, N., Steinmetz, KA. and Potter, JD.1996. Impact of environment factors on product quality of greenhouse vegetables for fresh consumption. *J. Plant Sci.* 24 (3): 227-247.

Lowry, O.H., Rosenberg, N.J., Ferr, A.L. and Randal, R.L. 1951. Protein measurement with folin-phenol reagent. *J. Biol.Chem.* 193:265-275.

Mahadevan, A. and Sridhar, R., 1982. Methods in Physiological Plant Pathology. Sivakami Publication, Madras.pp.127-129.

Malik, C.P. and Singh, M.B., 1980. Plant Enzymology and Histo Enzymology. Kalyani Publishers, New Delhi.pp.68-69.

Moore, S. and Stein, W.H., 1948. Photometric ninhydrin method for use in the chromatography of amino acids. *J. Biol.Chem.* 176:367-388.

Nelson, N., 1944. A photometric adaptation of the somogyi method for the determination of glucose. *J. Biol.Chem.* 153:375-380.

Okoye, Z.S.C., 1992. Biochemical aspects of nutrition. Prentice Hall of Indian private limited, New Delhi.p.69.

Saxena, K.P., 1970. Economic Botany. In: Vegetables. Prakashan Kendra Publication, Bangalore. p.9.

Srilakshmi, B. 1999. Food Science, 4th edition, New Age International Ltd. Publishers, New Delhi.p.7.

24

Phytochemical Screening and Free Radical Scavenging Activity of Selected Medicinal Plants

Gomathi, R. and Usha, K.

1. INTRODUCTION

Medicinal plants are nature's priceless gift to human. They have been used all over the world for the treatment and prevention of various ailments particularly in developing countries where infectious diseases are endemic and modern health facilities and services are inadequate. Medicinal principles are present in different parts of the plants such as root, stem, bark, heartwood, leaf, flower, fruit or plant exudates (Thenmozhi *et al.*, 2010). Phytochemicals from medicinal plants serve as lead compounds in drug discovery and design (Sofowora, 2006).

Free radicals are chemical species which contain one or more unpaired electrons, due to which they are highly metastable and cause damage to other molecules by extracting electrons from them in order to attain stability (Chakraborthy and Ghorpade, 2010). Hyper-physiological burden of free radicals causes imbalance in homeostatic phenomena between oxidants and antioxidants in the body. This imbalance leads to oxidative stress that is being suggested as the root cause of aging and various diseases like diabetes, cancer, and liver damage (in severe conditions like cirrhosis and hepatitis) (Wagh *et al.*, 2010).

DPPH is widely used to evaluate the free radical scavenging effects of various antioxidant substances and polyhydroxy aromatic compounds. It is a relatively stable free radical. The assay is based on the measurement of the scavenging ability of antioxidants towards the stable radical DPPH. DPPH radical reacts with suitable reducing agent. The electrons become paired off and solution loses color stoichiometrically depending on the number of electrons taken up (Pari *et al.*, 2011).

Nelumbo nucifera Garetn, namely Lotus, is a perennial aquatic plant grown and consumed throughout Asia. All parts of *Nelumbo nucifera* have been used in oriental medicine for various medicinal purposes (Ho *et al.*, 2010).

Onions and garlic have long been used for large range of purposes including medicine, flavorings, condiment, and the treatment of common ailments as folk medicine. Recent research has now reported that phenolic phytochemicals from onion have high antioxidant activity and blood glucose lowering effect in alloxan-induced diabetic rat (Kim *et al.*, 2011). Hence the present investigation was carried on *Nelumbo nucifera* flowers and *Allium cepa* bulbs.

2. MATERIALS AND METHODS

2.1 Collection and preparation of the plant samples

Fresh flowers of *Nelumbo nucifera* and fresh bulbs of *Allium cepa* were taken for this study. The plant samples were collected from Coimbatore District, Tamil Nadu, India. They were washed under running tap water to remove the surface pollutants and the different parts of the plant materials were separated mechanically. The separated parts were air dried under shade. The dried sample were powdered and used for further studies.

2.2 Chemicals

Ammonium hydroxide, methanol, Fehling's solution, DPPH (Diphenyl-2-picryl hydrazyl radical), acetic anhydride, ferric chloride, hydrochloric acid, ammonia, sulphuric acid were obtained from Loba and Merck. The solvents used were of analytical grade.

2.3 Extraction of Plant material

20 grams of powdered samples were weighed and wrapped separately with whatmann No.1 filter paper and extracted using different solvents such as benzene, chloroform, ethyl acetate, methanol and petroleum ether in soxhlet apparatus. The different solvent extracts were concentrated by rotary vacuum evaporator and then air dried. The dried extract obtained with each solvent was weighed. The extract obtained was used for the assessment of free radical scavenging assay and for further analysis.

2.4 Qualitative Phytochemical analysis

2.4.1 Test for alkaloids

50 mg of solvent free extract was stirred with one ml of dilute hydrochloric acid and filtered. The filtrate was tested for alkaloids.

Mayer's Test: To the filtrate, a drop of Mayer's reagent was added along the sides of the test tube. A white precipitate indicates the test as positive.

2.4.2 Test for flavonoids

Alkaline reagent test: Two ml of aqueous solution of the extract was treated with 1 ml of 10% ammonium hydroxide solution. Yellow fluorescence indicated the presence of flavonoids.

2.4.3 Test for saponins

50 mg of the plant extract was ground with 3 ml of distilled water and diluted with the same made upto 20 ml. The suspension was shaken in a graduated cylinder. After 15 minutes, a two cm layer of foam indicated the presence of saponins.

2.4.4 Test for phenols

Ferric chloride test: 50 mg of the sample was dissolved in 5 ml of distilled water. To this few drops of neutral 5% ferric chloride solution was added. A dark green colour indicates the presence of phenolic compounds.

2.4.5 Test for glycosides

50 mg of the plant extract was hydrolysed with concentrated hydrochloric acid for 2 hours on a water bath, filtered and the hydrolysate was subjected to the following test.

Borntrager's test: From the filtered hydrolysate, 3 ml of chloroform layer was separated and 2 ml of 10% ammonia solution was added to it. Pink colour indicates the presence of glycosides.

2.4.6 Test for carbohydrates

To 0.5ml of the extract of the plant sample, 1ml of water and 5-8 drops of Fehling's solution was added and observed for brick red precipitate.

2.4.7 Test for tannins

One ml of water and 1-2 drops of ferric chloride solution was added to 1 ml of extract of the plant sample. Blue colour was observed for gallic tannins and green black for catecholic tannins.

2.4.8 Test for steroids

Libermann-Burchard reaction: 4mg of the plant extract was treated with 0.5 ml of acetic anhydride and 0.5ml of chloroform. Then concentrated sulphuric acid was added slowly and green bluish colour for steroids was observed.

2.4.9 Test for terpenoids

4 mg of the extract was treated with 0.5 ml of acetic anhydride and 0.5ml of chloroform. Concentrated sulphuric acid was added slowly along the sides of the test tube. Red violet colour was observed for terpenoids.

2.5 Quantification Assay

2.5.1 DPPH radical scavenging activity

DPPH is widely used to evaluate the free radical scavenging effects of various antioxidant substances and polyhydroxy aromatic compounds. It is a relatively stable free radical. The assay is based on the measurement of the scavenging ability of antioxidants towards the stable radical DPPH (Pari *et al.*, 2011).

Sample extracts at various concentrations were taken and the volume was adjusted to 100 μL with methanol. 3 ml of 1mM DPPH in methanol was added to the aliquots of samples and shaken vigorously. DPPH solution with methanol was used as a positive control and methanol acted as a control. When DPPH reacts with antioxidant, the sample of DPPH was reduced and the color changed from deep violet to light yellow. This was measured at 517 nm. Radical scavenging activity of the samples was expressed as IC50 which is the concentration of the sample required to inhibit 50 % of DPPH˙ concentration. The percentage antioxidant activity was calculated by the following formula.

$$\text{Scavenging activity } (\%) = \frac{A_{517}\,(\text{control}) - A_{517}\,(\text{sample})}{A_{517}\,(\text{control})} \times 100$$

3. RESULTS AND DISCUSSION

3.1. Preliminary phytochemical screening

Preliminary phytochemical analysis was carried out in *Nelumbo nucifera* flowers and *Allium cepa* bulbs. In *Nelumbo nucifera* flowers, all the solvent extracts were found to contain various phytochemical constituents. Benzene extract showed the presence of alkaloids, amino acids, carbohydrates, steroids, glycosides and steroids. Alkaloids, anthocyanin, glycosides and phenols were present in chloroform extract. In ethyl acetate extract alkaloids, flavonoids, glycosides, steroids and phenols were present. Methanolic extract showed the presence of alkaloids, amino acids, anthocyanin, carbohydrates, flavonoids, glycosides, saponins, steroids, terpenoids and phenols. Alkaloids, carbohydrates, flavonoids, glycosides and steroids were present in peteroleum ether extract (Table1).

Allium cepa bulbs were evaluated for the presence of various phytochemical constituents. Benzene extract showed the presence of alkaloids, amino acids, anthocyanin, flavonoids, glycosides, tannins, phenols and fixed oils and fats. Alkaloids, flavonoids, glycosides, saponins, tannins and terpenoids were present in chloroform extract. Ethyl acetate extract showed the presence of alkaloids, anthocyanin, flavonoids, glycosides, saponins, fixed oils and fats. Alkaloids, amino acids, carbohydrates, flavonoids, glycosides, saponins, steroids, tannins, phenols, fixed oils and fats were present in methanloic extract. Alkaloids, anthocyanin, flavonoids, glycosides, tannins, terpenoids, phenols, fixed oils and fats were present in petroleum ether extract.

All the solvent extracts were found to contain various phytochemical constituents. When compared with different solvent extracts, methanolic extract was found to contain major active phytoconstitutents such as alkaloids, amino acids, carbohydrates, flavonoids, glycosides, saponins, steroids, tannins, phenols, fixed oils and fats in both the plants.

Phytochemical and therapeutic profile of sacred lotus (*Nelumbo nucifera*) was reported by Pulok *et al.* (2009). The extracts of rhizomes, seeds, flowers and leaves of sacred lotus have varied therapeutic potential. Several bioactive compounds have been derived from these plant parts belonging to different chemical groups including alkaloids, flavonoids, glycosides, triterpenoid, and vitamins.

Rao *et al.* (2005) stated that the preliminary phytochemical screening of ethanolic extract of *Nelumbo nucifera* flowers gave positive tests for alkaliods, flavoniods, lipids, steroids and triterpeniods, coumarin and aryl esters, proteins, fatty oils, tanins and resins.

Poluru *et al.* (2011) reported that the preliminary phytochemical screening of aqueous flower and rhizome extract of *Nelumbo nucifera* showed the presence of alkaloid, carbohydrate, flavoniods, terpenoids, proteins and amino acids.

Bang *et al.* (2010) reported that onion (*Allium cepa*) is rich in two phytochemical groups which have perceived health benefits to human being. These are the flavoniods and the alk(en)yl cysteine sulphoxides (ACSOs). The flavonols are often found concentrated in the outside skin of onion.

Chitra *et al.* (2009) on preliminary phytochemical investigation and wound healing activity of *Allium cepa Linn* (*Liliaceae*) reported that chloroform extract of bulbs of the plant possess alkaloids while ethanolic and aqueous extracts showed the presence of phytochemical constituents such as flavonoids, tannins, glycosides, carbohydrates and proteins.

Flavoniods, the plant-derived antioxidants, are defined as nonnutritive dietary components that are abundant in food. Potential use of flavonoids as functional food materials might be expanded because of their useful physiological properties such as prevention of cardiovascular disease and cancer (Boyle *et al.*, 2006).

Polyphenols, especially phenolic acids and flavonols have been considered as active components in the prevention of cardiovascular diseases, cancers, neurodegenerative diseases and diabetes. Epidemiological and *in vitro* studies strongly suggest that foods containing phytochemical antioxidants have potentially protective effects against many diseases including cancer, diabetes and cardiovascular diseases (Jiang *et al.*, 2010). Phytochemicals such as carotenoids, tocopherols, ascorbates and phenols present in plants are natural antioxidants and have an important role in health care systems (Prakash *et al.*, 2007).

Plants containing flavonoids have been reported to possess strong antioxidant properties. Flavonoids are a group of polyphenolic compounds diverse in chemical structure and chararcteristics. They are widely distributed in foods of plant origin such as vegetables, fruits, tea and wine. Scavenging of free radicals seems to play a considerable part in the antioxidant activity of flavoniod compounds (Pari and Amudha, 2011). Thus the preliminary screening test indicates the presence of the above mentioned bioactive compounds. Subsequently it may be used for the preparation of drug in a systematic way which may lead to the cure of many ailments in the future.

Table 1: Phytochemical screening of *Nelumbo nucifera* flowers

Phytochemical Constituents	Benzene	Chloroform	Ethyl acetate	Methanol	Petroleum ether
Alkaloids	+	+	+	+	+
Anthraquinons	-	-	-	-	-
Amino acids	+	-	-	+	-
Anthocyanin	-	+	-	+	-
Carbohydrates	+	-	-	+	+
Flavonoids	-	-	+	+	+
Glycosides	+	+	+	+	+
Saponins	-	-	-	+	-
Steroids	+	-	+	+	+
Tannins	-	-	-	-	-
Terpenoids	-	-	-	+	-
Phenols	-	+	+	+	-
Fixed oils and fats	-	-	-	-	-

(+): Presence of phytochemical compound, (-): Absence of phytochemical compound

Table 2: Phytochemical screening of *Allium cepa* bulbs

Phytochemical Constituents	Benzene	Chloroform	Ethyl acetate	Methanol	Petroleum ether
Alkaloids	+	+	+	+	+
Anthraquinons	-	-	-	-	-
Amino acids	+	-	-	+	-
Anthocyanin	+	-	+	-	+
Carbohydrates	-	-	-	+	-
Flavonoids	+	+	+	+	+
Glycosides	+	+	+	+	+
Saponins	-	+	+	+	-
Steroids	-	-	-	+	-
Tannins	+	+	-	+	+
Terpenoids	-	+	-	-	+
Phenols	+	-	-	+	+
Fixed oils and fats	+	-	+	+	+

(+): Presence of phytochemical compound, (-): Absence of phytochemical compound

3.2 Quantification Assay

3.2.1 DPPH radical scavenging activity

DPPH is widely used to evaluate the free radical scavenging effects of various antioxidant substances and polyhydroxy aromatic compounds. It is a relatively stable free radical. The assay is based on the measurement of the scavenging ability of antioxidants towards the stable radical DPPH. DPPH radical reacts with suitable reducing agent. The electrons become paired off and solution loses color stoichiometrically depending on the number of electrons taken up (Pari *et al.*, 2011).

2, 2'-diphenylpicrylhydarzyl (DPPH) has been used extensively as a free radical to evaluate reducing substances. In the present study, the antioxidant activity of methanolic extracts of *Nelumbo nucifera* flowers (MNnF) and *Allium cepa* bulbs (MAcB) were evaluated by DPPH (2, 2'-diphenylpicrylhydarzyl) radical scavenging activity

The DPPH radical scavenging activity was carried out for different concentrations of MNnF and MAcB. The results of DPPH assay were expressed in IC50 values. The lower value of IC50 indicates a higher antioxidant activity. Among the two plants extract, MNnF showed better DPPH radical scavenging activities compared to MAcB (Fig. 1). Butylated hydroxyltoluene (BHT) was used as reference compound.

MNnF and MAcB showed the dose dependent DPPH radical scavenging activity. From the graph, the 50 percent effective/inhibitory concentrations (IC_{50}) of MNnF, MAcB and BHT were found to be 55μg, 70μg and 85μg respectively.

Fig. 1

The free radicals are the byproducts of aerobic metabolism and are produced by endogenous sources such as mitochondrial leak, respiratory burst, enzyme reactions and auto-oxidant reactions and environmental sources such as cigarette smoke, pollutants, ultraviolet light, ionizing radiation and xenobiotics (Kalaivani *et al.*, 2010).

According to Kim *et al.* (2011) as concentration of extracts increased, DPPH free radical scavenging activity was increased. At 100μg/ml extract concentration, *Nelumbo nucifera's* leaf, flower and seed extract showed 78%, 78% and 80% anti-oxidation effect respectively.

4. CONCLUSION

Medicinal plants have curative properties due to the presence of various complex chemical substances of different composition, which are found as secondary plant metabolites in one or more parts of these plants. In the present study *Nelumbo nucifera* flowers and *Allium cepa* bulbs were found to contain various phytochemical constituents. Phytopharmaceuticals are gaining importance in allopathic as well as traditional medicine owing to their non-addictive and less toxic nature. The lower value of IC50 indicates a higher antioxidant activity. Among the two plant extracts MNnF showed better DPPH radical scavenging activities compared to MAcB. Free-radical scavengers are antioxidants which provide protection to living organisms from damage caused by uncontrolled production of reactive oxygen species and subsequent lipid peroxidation, protein damage and DNA strand breaking. Novel antioxidants may offer an effective and safe means of counteracting some of the problems and blostering the body's defense against free radicals and cardiovascular diseases.

REFERENCES

Bang,M,A., and Kim,H,A. (2010), Dietary Supplementation of Onion Inhibits Diethylnitrosamine-induced Rat Hepatocellular Carcinogenesis, *Food Sci.Biotechnol.* 19(1): 77-82.

Boyle S.P., Dobson, V.L., Duthie, S.J., Hinselwood,D.C., Kyle,J.A., Collins,A.R. (2006), Bioavailability and efficiency of rutin as antioxidant: a human supplementation study, *Eur J Clin Nutr*, 54:774-782.

Chakraborthy,G.S. and Ghorpade,P.M. (2010), Free radical scavenging activity of *Abutilon indicum*(L.) sweet stem extract, *Int.J.Chem.Tech Research*, 2 (1): 526-531.

Chitra,S., Patil, M.B., Kumar, R. and Patil, S. (2009), Preliminary phytochemical investigation and wound healing activity of *Allium cepa* Linn. (Liliaceae), *International Journal of Pharmacy and Pharmaceutical Sciences*, 2(2), 167-175.

Ho,H,H., Hsu,L,S., Chan,K,C., Chen,H,M., Wu,C,H., and Wang, C,J. (2010), Extract from the leaf of nucifera reduced the development of atherosclerosis via inhibition of vascular smooth muscle cell proliferation and migration, *Food Chem Toxicol*, 48:159-168.

Jiang,Y., Ng,T.B., Liu,Z., Wang,C., Li,N., Qiao,W and Liu,F. (2010), Immunomodulatory and anti-HIV-1 enzyme activities of antioxidants components from lotus (*Nelumbo nucifera* Gaertn) rhizome, Bioscience Reports Immediate Publication, 10.1042/BSR20100062.

Kalaivani,T., Rajasekaran,C., Suthindhiran,K. and Lazar Mathew. (2010), Free radical scavenging, cytotoxic and hemolytic activities from leaves of *Acacia nilotica*(l.) wild Ex.Delile., subsp., indica (Benth.) Brenan, eCam advance published, 1-8.

Kim,M,H., Jo,S,H., Jang,H,D., Lee,M,S., and Kwon,Y,I. (2011), Antioxidant Activity and α-Glucosidase Inhibitory Potential of Onion (*Allium cepa* L.) Extracts, *Food Sci.Biotechnol.* 19(1): 159-164.

Pari,L., and Amudha K., (2011), Antioxidant effect of naringin on nickel-induced toxicity in rats: an *in vivo* and *in vitro* study, *IJPSR*, 2(1): 151-158.

Poluru, R., Sekar, S. and Kumar, S. (2011), A comparative study on the antidiabetic effect of *Nelumbo nucifera* and glimerpiride in streptozotocin induced diabetic rats, 2(2): 63-68.

Prakash, D., Suri, S., Upadhya, G. and Singh, B.N. (2007), Total phenol, antioxidant and free radical scavenging activities of some medicinal plants, *International Journal of Food Sciences and Nutrition,* 58: 18-28.

Pulok K. Mukherjee, Debajyoti Mukherjee, Amal K. Maji, Rai, S. and Michael Heinrich. (2009), The sacred lotus *(Nelumbo nucifera)*– phytochemical and therapeutic profile, *Journal of Pharmacy and Pharmacology*, 61(4): 407-422.

Rao, G.M.M., Pushpangandan, P. and Shirwaikar, A. (2005), Hepatoprotective activity of *Nelumbo nucifera Gaertn* flower: an Ethnopharmacological study, *Acta Pharmaceutica* Turica, 47: 79-88.

Sofowora,A. (2006), Medicinal Plants and Traditional Medicine in Africa, Specturm Books, Ibadan, 150-165.

Thenmozhi,M., Sivaraj,R. and Yadav,K.R.(2010), Acomparative phytochemical analysis of *Alstoni scholaris, Lawsonia inermis, Ervatamia divartica* and *Asparagus racemosus, Int J Pharma. Research and Development.*, 2(9): 86-91.

Wagh,S.S. and Shinde,G.B.(2010), Antioxidant and hepatoprotective activity of *Tridax procumbens*, Linn. against paracetamol induced hepatotoxicity in male albino rats, *Advanced Studies in Biology.*, 2 (3); 105-112.

25

Studies on the Rare Medicinal Plants of Karaikal District, U.T of Puducherry

Kathirvelu Sambandan and Narayanasamy Dhatchanamoorthy

1. INTRODUCTION

The medicinal plants have been used by humans from the pre-historical times. Traditional medicine based on herbal remedies has always played a key role in the health systems of many countries. In India, the native people are exploiting a variety of herbals for effective curing of various ailments. Medicinal plants are widely used by all sections of the population and country is richly endowed with a wide variety of plants of medicinal value, which represents the great national resource (Alok, 1991).

It is estimated that at least 70 percent of population relay on herbal medicines for primary health care in our country (Holley and Williams, 1996). In India, different classical medicinal systems such as Ayurveda, Siddha and Unani are being practiced in the country and in addition to innumerable local folk medicinal traditions exist. More than 8000 species of wild plants are known to be used in India for the treatment of various health problems (Ved *et al.*, 2000). This rich medicinal wealth is mainly distributed in two biodiversity hotspots that are Eastern Himalayas and Western Ghats. Numerous studies have pointed out that many drugs that are used in commerce have come from folk-use and use of plants by indigenous cultures (Anon, 1994). Medicinal plants form a heterogeneous group in growth, habit, distribution, reproduction, phenology and their ecological requirements. About fifty drugs have been discovered from ethnobotanical leads by translating folk knowledge into new pharmaceuticals (Cox, 1994). Some examples of medicinal plant from the Asia-Pacific region are of species such as *Rauvolfia*, *Hyoscyamus*, *Cassia*, *Atropa*, *Podophyllum*, *Psoralea*, *Catharanthus*, *Papaver,* etc. However, relatively few medicinal and aromatic plant species have been brought into cultivation worldwide and most of the medicinal species continue to be harvested from their native habitats (Gupta and Chadha, 1995; Salleh *et al.*, 1997; Gautam *et al.*, 1998).

Very little is known about the distribution, abundance, ecology, cultivation and genetic diversity of majority of the medicinal plants, although some efforts have started in recent years (Chadha and Gupta, 1995; Chandel *et al.*, 1996; Kumar *et al.*, 2000; Paisooksantivatana *et al.*, 2001), including the use of molecular markers (Sharma *et al.*, 2000; Natesh, 2000). The collection and processing of medicinal plants in India provide employment and income opportunities for a large number of people in rural areas. With the large increase in domestic and international demand, there has been a sharp increase in collection of medicinal plants from the wild conditions. This has put extreme pressure on wild resources, and many species face threat of extinction. On the other side, Habitat destruction is the major threat for the survival of medicinal plants. Proper documentation of indigenous and ethnobotanical knowledge in each country will help to establish the base line data and to plan for conservation programmes (Anon., 1994). In the present study is an attempt to record the rare medicinal plants of Karaikal district and to document their traditional medical practices.

2. MATERIALS AND METHODS

Karaikal district of Puducherry state lies between 10°49' and 11° 01' N latitude and 78°43' and 79°52' E Longitude. It is one of four erstwhile French establishments of the Union Territory of Puducherry with long coastal border of Bay of Bengal. Topographically, Karaikal district forms a broad plain, parallel to the coast and gradually raising to the western side. As a part of fertile delta, the district is completely covered with distributaries of river Kaveri. Climate of Karaikal is very warm and dry throughout the year. The average annual rainfall varies from 827 mm to 2100 mm (during last ten years) and it is received both from Southwest monsoon (June to September) and Northeast monsoon (October to December). Throughout the year, the coastal areas of Karaikal district show higher relative humidity (average value 78%). Study area composed of variety of vegetation which covers coastal, aquatic, agricultural, riparian, waste land, village and urban areas of the Karaikal district. Floristic survey of rare medicinal plants was carried out in all the communes of Karaikal district (Fig 1). Intensive exploration trips were undertaken for about three years period (2009-2011). Field trips were made at frequent intervals especially to obtain a through collection of ephemerals and staggered over different seasons, namely, post-monsoon (January to March), summer (April to June), pre-monsoon (July to September) and monsoon (October to December). The fresh specimens are soaked in 1% formaldehyde and processed in to Herbarium specimens were followed the standard methods (Henry *et al.*, 1989; Forsberg & Sachet, 1965). These herbarium sheets were mounted and deposited in the herbarium of plant science laboratory of Avvaiyar Government College for Women, Karaikal. Each specimen was be carefully examined in natural conditions, identify taxonomically using publications of Gamble and Fischer (1915-1936), Henry *et al.*, (1989) and Matthew (1982; 1983; 1988; 1991 and 1999) as well as use of various floras, monographs and revisions, provisional identification were made. Botanical nomenclature followed in the text conforms to the Flora of Tamil Nadu by Nair and Henry (1983) and Henry *et al.*, (1987, 1989). Identification was confirmed after matching the herbarium specimens with authentication or type specimens in Madras Herbarium (MH), Coimbatore and Central National Herbarium, Howrah (CAL). The nomenclature of the plants has been up-to-date by strict application of the rules as given in the International Code of Botanical Nomenclature (Voss *et al.*, 1983). Further, every attempt was made to check the identity of plants by the way of type method and evaluate the taxonomic status of the taxa concerned.

The ethno botanical survey was carried out among local population in their residential areas of different villages in the district. The field visit was conducted several times during year 2010 to 2011 in the study area. Ethnobotanical data was collected according to the methodology suggested by Jain (1987) through questionnaire, interviews and formal and informal discussions among villagers in their local language. Ethno medicinal values of plants were ascertained in consultation with village people using various methods (Jain, 1987). In this study, questionnaire was allowed descriptive response on the plant prescribed, such as part of the plant used, medicinal uses, and detailed information about mode of preparation and type of usages. The data was collected from 41 informants between the ages of 34 to 68 of both men and women in the study area. Some of them were accompanied us to the field where they showed us plants that are used as medicines. Villagers were also interviewed for preparation and administration of drugs. For local uses, a cross section of village populations were interviewed and cross interviewed.

3. RESULTS AND DISCUSSION

In the study, data were tabulated and arranged in the sequence of serial number, botanical name, family, vernacular name and ethno medicinal uses (Table-1). Uses of twenty rare medicinal plants reported from Karaikal District, U T. of Puducherry. Perhaps, this is first hand investigation of rare medicinal plants among the villagers in the district.

Table 1: Rare medicinal plants occurring in Karaikal district

S. No	Plants Name	Family	Habit
1.	***Ammania octandra*** L.f.	Lythraceae	Herb
2.	***Anisomeles malabarica*** (L.) R. Br. ex Sims	Lamiaceae	Under shrub
3.	***Cardiospermum canescens*** Wall.	Sapindaceae	Climber
4.	***Centella asiatica*** (L.) Urban	Apiaceae	Herb
5.	***Cleome aspera*** Koen ex. DC.	Cleomeaceae	Herb
6.	***Clerodendrum phlomidis*** L.f.	Verbenaceae	Shrub
7.	***Derris trifoliata*** Lour.	Fabaceae	Climber
8.	***Glinus lotoides*** L.	Molluginaceae	Herb
9.	***Grangea maderaspatana*** (L.) Poir.	Asteraceae	Herb
10.	***Hibiscus panduriformis*** Burm.	Malvaceae	Under shrub
11.	***Indigofera aspalathoides*** Vahl ex DC.	Fabaceae	Under shrub
12.	***Indigofera colutea*** (Buem. F.) Merr.	Fabaceae	Herb
13.	***Mitragyna parvifolia*** (Roxb.) Korth.	Rubiaceae	Under shrub
14.	***Phyllanthus rotundifolius*** Klein ex Willd.	Euphorbiaceae	Shrub
15.	***Polygonum plebeium*** R. Br.	Polygonaceae	Prostrate herb
16.	***Sida spinosa*** L.	Malvaceae	Under shrub
17.	***Tephrosia pumila*** (Lam.) Pers.	Fabaceae	Climber
18.	***Urena lobata*** L. subsp. *sinuata* (L.) Borssum	Malvaceae	Under shrub
19.	***Waltheria indica*** L.	Sterculiaceae	Herb
20.	***Withania somnifera*** (L.) Dunal	Solanaceae	Under shrub

Ammania octandra L.f.
Habit: Herb
Habitat: In wet places, marshy places and paddy fields
Uses: Leaves are used for the ringworm and other parasitic skin infections
Fl. & Fr.: Nov. - March.

Anisomeles malabarica (L.) R. Br. ex Sims
Habit: Undershrub
Habitat: In waste places
Uses: Infusion of the leaves has been found to be beneficial in peptic ulcers and fever, accompanying teething in children. Total plant is used as antidote to scorpion sting and snake-bite. Paste of the leaves applied externally to cure eczema and chronic wounds.
Local Name: Peimiratti

Cardiospermum canescens Wall.
Habit: Climber
Habitat: In scrub jungles and paddy fields
Uses: Leaf paste is applied for curing abdominal pain. Roots are used for constipation and urinary disorders. It is also used for rheumatism and nervous diseases.
Local Name: Kaattu mudakkaruthaan
Fl. & Fr.: Dec. – Feb.

Centella asiatica (L.) Urban
Habit: Perennial prostrate herb
Habitat: In moist localities, shade places near of paddy fields
These plants used for Diuretic and Leprosy. Leaf Paste is used to cure dysentery. Plant powder used to improve memory power in children
Local name: Vallarai
Fl. & Fr.: Nov. – Mar.

Cleome aspera Koen ex. DC.
Habit: Herb
Habitat: In sandy localities and wastelands
Uses: Leaves are used for diabetes. Leaf paste is used to cure skin rashes and other skin disorders.
Local Name: Siru kadugu

Clerodendrum phlomidis L.f.
Habit: Shrub
Habitat: In stream margins and thickets
Uses: Root is used as a demulcent in gonorrhea
Local Name: Thazhuthaazhai
Fl. & Fr.: Sep. – March.

Derris trifoliata Lour.
Habit: Liana
Habitat: In marshy localities and thickets
Uses: Bark is used as fish poison. It is useful in rheumatism, dysentery, antispasmodic agent and counter-irritant.
Fl. & Fr.: Oct. – Nov.

Glinus lotoides L.
Habit: Prostrate herb
Habitat: In waste places, roadsides and dry fields
Uses: The plant is useful in abdominal disorders. The tender shoots are eaten as pot herb.
Local Name: Siruseruppadai
Fl. & Fr.: April – Aug.

Grangea maderaspatana (L.) Poir.
Habit: Herb
Habitat: In wasteland and near of paddy fields
Uses: Leaves are used as a valuable medicine for obstructed menses, hysteria, stomach pain and antispasmodic properties. It is used for in preparing antiseptic and antidote formulations.

Local Name: Maasipathri
Fl. & Fr.: Jan. – April.

Hibiscus panduriformis Burm.
Habit: Undershrub
Habitat: In near of paddy fields, wastelands on roadsides.
Uses: Oil from the seed is used for the skin dryness whole plants are used as leafy vegetable for villagers.
Fl. & Fr.: Throughout the year.

Indigofera aspalathoides Vahl ex DC.
Habit: Herb
Habitat: In sandy localities
Uses: Roots are chewed as a remedy for toothache. Leaf, flower and tender shoots are used as cooling agent for body. Decoction of leaves, flowers and shoots are used for ulcers and leprosy.
Leaves are also used for abscesses and cancerous tumors.
Local Name: Sivanaar vaembu
Fl. & Fr.: Nov. – April.

Indigofera colutea (Buem. F.) Merr.
Habit: Herb
Habitat: In scrubs jungles and roadsides
Uses: Plant used for anti-plasmodial drug and also treatment of stomach pain.
Fl. & Fr.: Oct. – March.

Mitragyna parvifolia (Roxb.) Korth.
Habit: Tree
Habitat: In stream margins and river banks
Uses: Bark juice is used as an antidote to snake bite and wash as a disinfectant. Leaf paste is used to give relief for colic pains. Flower buds are powdered and filled in the cavities of carious teeth to relief pain and swelling. Fruit powder is boiled in water and goggled to cure gingivitis. It is also used for wounds, skin diseases and leprosy.
Local Name: Neer kadambu
Fl.: March – July. Fr.: June – Aug.

Phyllanthus rotundifolius Klein ex Willd.
Habit: Herb
Habitat: In sandy localities
Uses: Leaf sap is applied for wounds. Leaf juice is used for eye Infections. Root decoction is used for venereal diseases.
Fl. & Fr.: Nov. – July

Polygonum plebeium R. Br.
Habit: Herb
Habitat: In dry localities and wastelands
Uses: Dried powder of whole plant is taken internally to cure pneumonia. Dried root powder is given for bowel complaints. The plant is useful in abdominal disorders. The tender shoots are eaten as pot herb.
Fl. & Fr.: Nov. - May.

Sida spinosa L.
Habit: Undershrub
Habitat: In wastelands
Uses: Root and root bark are used as a demulcent in irritability of the bladder and in gonorrhoea, tonic prepared from the root is used in fever.
Local Name: Arival manai poondu
Fl. & Fr.: Sep. – Feb.

Tephrosia pumila (Lam.) Pers.
Habit: Herb
Habitat: In wastelands and moist localities
Uses: Plant is used for broad spectrum antifungal and antibacterial agent
Fl.: Oct. - May. Fr.: Dec. - June.

Urena lobata L. subsp. **sinuata** (L.) Borssum
Habit: Undershrub
Habitat: Near of paddy fields, stream margins and roadsides
Uses: Root is useful against rheumatic pain. Seeds and bark of this plant are used for diarrhea
And skin diseases. Total plant is used for diabetes.
Local Name: Ottatthi
Fl. & Fr.: Oct. – April.

Waltheria indiica L.
Habit: Herb
Habitat: In wastelands and wasteplaces
Uses: Root is chewed to control internal haemorrhages. A decoction of roots is used to induce fecundity in women. The plant powder is also used against cough. Flowers and root-barks used for fungal infections.
Local Name: Chembudu
Fl. & Fr.: Throughout the year.
Withania somnifera (L.) Dunal
Habit: Undershrub
Habitat: In roadsides

Uses: It is used in cases of debility from old age. Roots are used for cough, rheumatism and as a sedative. Leaf paste applied to lesions, painful swelling and sore eyes.
Local Name: Amukkira
Fl. & Fr.: April – Dec.

During the survey, 20 species of rare medicinal plant species belonging to 20 genera and 14 families have been collected in this region. *Fabaceae* is the dominant family with 4 rare medicinal plant species followed by *Malvaceae* (3) and *Euphorbiaceae* (2). Notable rare medicinal plants, especially, *Derris trifoliata, Mitragyna parvifolia*, *Grangea maderaspatana*, *Clerodendrum phlomidis*, *Sida spinosa*, *Phyllanthus rotundifolius* and *Cardiospermum canescens* are recorded. These medicinal plants and its healing and curing properties are well adapted themselves in the district. Keeping the importance of plants in consideration, large numbers of commercially important medicinal plant species are over-exploited by persons involved in the trade. Lack of sustainable harvesting methods, inadequate knowledge about management and lack of financial resources are the main causes of over-exploitation, indiscriminate collection of some commercially important species, and the populations of valuable species are decline in the district.

The study provides sufficient ground that the traditional medicinal practice using native medicinal plants is alive well functioning in the study area. Many villages in the district use wild plant parts for the primary healthcare, due to belief in its effectiveness as a medicine. The treatment of disease with plants and its products also causes little site effects and are cost effective too.

The increased availability of medicinal plant products is expected to reduce the burden on the naturally occurring medicinal plants and will help in sustainable conservation of these important natural resources. Consequently, cultivation of these plants is urgently needed to ensure their availability to the industry as well as to people associated with traditional system of medicine. If timely steps are not taken for their conservation, cultivation and mass propagation, they may be lost from the natural vegetation forever. *In situ* conservation of these resources alone cannot meet the ever increasing demand of pharmaceutical industry. It is, therefore, inevitable to develop cultural practices and propagate these plants in suitable agro-climatic regions. Commercial cultivation will put a check on the continued exploitation from wild sources and serve as an effective means to conserve the rare floristic wealth and genetic diversity. It is necessary to initiate systematic cultivation of medicinal plants in order to conserve biodiversity and protect endangered species.

4. CONCLUSION

Traditional medicine based on herbal remedies has always played a key role in the health systems of many countries. In India, the native people are exploiting a variety of herbals for effective curing of various ailments. Living close to nature, traditional societies have acquired unique knowledge about the use of wild flora and fauna, most of which are unknown to the people who live away from such natural ecosystems. The present study found that rare medicinal plants are recorded during preparation of district flora is highly valuable for various medicinal uses. The study provides sufficient ground to believe that the traditional medicinal practice using native medicinal plants is alive well functioning in the study area.

During the survey, 20 species of rare medicinal plants belonging to 20 genera and 14 families have been collected in the Karaikal District of Puducherry region. *Fabaceae* is the dominant family with 4 rare medicinal plant species followed by *Malvaceae* (3) and *Euphorbiaceae* (2). Notable rare medicinal plants, especially, *Derris trifoliata*, *Mitragyna parvifolia*, *Grangea maderaspatana*, *Sida spinosa*, *Phyllanthus rotundifolius*, *Cardiospermum canescens* and its traditional uses are recorded. These medicinal plants and its healing and curing properties are well adapted themselves in the district. Lack of sustainable harvesting methods, inadequate knowledge about management and rapid industrialization are the main causes of destruction of valuable species in the district.

ACKNOWLEDGEMENT

The authors are grateful to the rural people in Karaikal district who whole heartily co-operated in sharing their knowledge and in helping the collection of the plant material for study. Author (K.S) grateful to the University Grants Commission (UGC) New Delhi for providing the financial assistance in the form of Major research Project and to the Principal and Head of the Department of Plant Science, Avvaiyar Government College for Women, Karaikal for the facilities and encouragement.

REFERENCES

Alok, S.K., 1991, Medicinal plants in India: Approaches to exploitation and conservation. In conservation of medicinal plants, Olayiwola Akerele, Vernon Hey wood and Hugh Synge (eds.) Cambridge, pp. 295-303.

Anonymous, 1994. Ethnobotany in the search for new drugs. Ciba Foundation Symposium 188, John Wiley and Sons, New York.

Chandel, K.P.S., Shukla, G., Sharma, N., 1996. Biodiversity in Medicinal and Aromatic Plants in India. National Bureau of Plant Genetic Resources, New Delhi, India.

Cox, P.A., 1994. The ethnobotanical approach to drug discovery: Strengths and limitations. Ethnobotany in the Search for New Drugs. Ciba Foundation Symposium 188, John Wiley and Sons. New York. pp. 25-41.

Fosberg, F.R., Sachet, M.H., 1965. Manual for tropical herbaria. Regnum Veg., Vol.39.

Gamble, J. S., Fischer, C.E.C., 1915-38. Flora of Presidency of Madras, London.

Gautam, P.L., Raina, R., Srivastava, U., Raychaudhari, S.P., Singh, B.B., 1998. Prospects of medicinal plants. Proceedings of UHF-IUFRO International workshops on prospects of medicinal plants. Indian Society of Plant Genetic Resources, New Delhi, India.

Gupta, R., Chadha, K.L., 1995. Medicinal and aromatic plants research in India. Advances in Horticulture: Medicinal and Aromatic Plants, Vol. 11. Malhotra Publishing House, New Delhi, pp. 1-43.

Henry, A,N., Chitra, V., Balakrishnan, N.P., 1989. Flora of Tamil Nadu, India, Series 1, Vol-3. Botanical Survey of India, Coimbatore, India.

Henry, A.N., Kumari, G.R., Chitra, V., 1987. Flora of Tamil Nadu, India Series-1, Vol-2, Botanical Survey of India, Coimbatore, India.

Holley, J., Williams, J.T., 1998. India uniquely positioned to capitalize on abundant medicinal plants heritage. *Diversity,* 12 (3): 35-37.

Jain, S.K., 1987. A Manual of Ethno-botany, Scientific Publishers, Jodhpur, India.

Kumar, S., Hassan, S.A., Dwivedi, S., Kukreja, A.K., Sharma, A., Singh, A.K., Sharma, S., Tewari, R., 2000. Proceedings of the National Seminar on the Frontiers of Research and Development in Medicinal Plants, September 2000. *Journal Medicinal and Aromatic Plant Sciences* Vol. 22/4A and Vol. 23/1A. Central Institute of Medicinal and Aromatic Plants (CIMAP). Lucknow, India.

Matthew, K.M., 1982. Illustrations on the Flora of Tamil Nadu Carnatic. Rapinat Herbarium, St' Josephs College, Trichirapalli, India.

Matthew, K.M., 1983. Flora of Tamil Nadu Carnatic, Vol **1-3** Rapinat Herbarium, St' Josephs College, Trichirapalli, India.

Matthew, K.M., 1988. Further Illustrations on the Flora of Tamil Nadu Carnatic. Rapinat Herbarium, St' Josephs College, Trichirapalli, India.

Matthew, K.M., 1991. An Excuration flora of Central Tamil Nadu Carnatic. Oxford and IBH Co, New Delhi, India.

Matthew, K.M., 1999. Flora of Palni Hills. Rapinat Herbarium, St' Josephs College, Trichirapalli, India.

Nair, N.C., Henry, A.N., 1983. Flora of Tamil Nadu, India Series-1, Vol-**1**,. Botanical Survey of India, Coimbatore, India.

Natesh, S., 2000. Biotechnology in the conservation of medicinal and aromatic plants. Biotechnology in Horticulture and Plantation Crops. K.L Chadha, P.N Ravindran and Leela Sahajram (eds), Malhotra Publishing House, New Delhi, India, pp. 548- 561.

Paisooksantivatana, Y., Kako, S., Seko, H., 2001. Isozyme polymorphism in Curcuma alismatifolia Gagnep. (Zingiberaceae) populations from Thailand. Scientia Horticulturae 88(4): 299-307.

Salleh, M.K., Natesh, S., Osman, A., Kadir, A.A., 1997. Medicinal and aromatic plants: Strategies and technologies for conservation. Proceedings of the Symposium State-of the- Art Strategies and Technologies for Conservation of Medicinal and Aromatic Plants. Kuala Lumpur, Malaysia, September 1997. Ministry of Science, Technology and Environment, Malaysia and Forest Research Institute, Malaysia.

Sharma, K.D., Singh, B.M., Sharma, T.R., Katoch, M., Guleria, S., 2000. Molecular analysis of variability in *Podophyllum hexandrum* Royle - An endangered medicinal herb of northwestern Himalaya. Plant Genetic Resources Newsletter (FAO/IPGRI) 124, 57-61.

Ved, D.K., Tandan, V., Ravikumar, K., Udyan, P.S., Utkarsh Ghate., 2000, Conservation assessment of medicinal plants (CAMP) in India (Unpublished). World conservation monitoring center (WCMC), 1992, Global Biodiversity: status of earth living resources. Chapmam and Hall, London.

Voss, E.G., 1983. International Code of Botanical Nomenclature. Regnum Veg., 111,1-80.

26

Assessment of Humoral Immune Response in Rabbits Fed with *C. occidentalis*

V. Ranganathan and S. Selvasubramanian

1. INTRODUCTION

Several herbs have been used traditionally to prevent and treat human and animal diseases. Recently, scientific evaluation of plants and preparations of plant origin medications have received more attention. *Cassia occidentalis* had been reported to be used as antidote of poison, blood purifier, expectorant, anti inflammatory agent and a remedy for the treatment of liver diseases (Ghani, 1920, Kabiruddin, 1951). *C occidentalis* was tested for antifeedant and insecticidal properties. It was studied successfully for its anti-parasitic (Tona *et al.*, 1999), hepatoprotective activity (Jafri *et al.*, 1999). Stimulation of the immune system is highly desirable for the treatment of immunodeficiency and infectious diseases. The anti-inflammatory and hepato protective effects of *C occidentalis* may relate to an influence of plant compounds on immunocompetent cells. The present study was undertaken to find out the effect of methanolic extract of *C occidentalis* leaves on the humoral immune response in rabbits.

2. MATERIALS AND METHODS

2.1 Extraction

The leaves of the plant was air dried and 150 grams of air dried powder was extracted with 400 ml of 80% methanol (v/v) using soxhlet apparatus and the extract was subjected to rotary evaporator for removal of the solvent. The paste form of extract was obtained using hot air oven and kept at room temperature for use.

2.2 Experimental animals used

Thirty six Newzealand white crossbred male rabbits weighing between 1000 to 1500 g obtained from the Department of Laboratory animal Medicine, TANUVAS, Madhavaram, Chennai-51, were used for the study. All the animals were given a balanced diet and maintained under identical environmental conditions.

2.3 Chemicals used

Immunostimulatory drug levamisole hydrochloride (Alved, Chennai) and immunosuppressive drug dexamethasone sodium (Merind, Mumbai) were used for this study.

2.4 Experimental design

Thirty six male Newzealand White rabbits were divided into six groups (I, II, III, IV, V, and VI) of six animals in each. Group I was the untreated control. Group II was dexamethasone sodium (Immunosuppressive) treated. Group III was Levamisole Hcl (Immuno stimulant) treated. Group IV was *C occidentalis* treated. Group V was levamisole and dexamethasone treated and Group VI was dexamethasone and *C occidentalis* treated group. Levamisole Hcl was given at 2.5 mg/kg subcutaneously thrice a week, dexamethasone sodium was given at the rate of 2mg/kg intra muscularly for seven days and *C occidentalis* was given @ 1000 mg/kg orally for 10 days. To assess the humoral immune response, estimations of antibody titre, total immunoglobulin concentration and haematology were carried out.

2.5 Assays

2.5.1 Estimation of antibody titre

All the eight groups of rabbits were immunized with typhoid-H antigen (King Institute, Guindy, Chennai) at the dose of one ml subcutaneously on 1st day and 15th day. Widal tube agglutination test was performed using widal test kits (Span Diagnostics Ltd.,) on days 0,7,14 and 21 for antibody titre.

2.5.2. Estimation of total immunoglobulin concentration

2.5.2.1. Ammonium sulphate Precipitation and dialysis method (Talwar and Gupta, 1983)

Five ml of saturated ammonium sulphate solution was added slowly to 5ml serum in a test tube to get half saturation of ammonium sulphate in the sample. Contents were mixed continuously by swirling the tube without agitating excessively. The mixture was allowed to stand for 5 minutes so that globulin precipitates. The material was centrifuges for 10 minutes at high speed. The precipitate was allowed to settle down at the bottom of the tube. The supernatant fluid (containing albumin) was poured off and the precipitate was redissolved in 1 ml distilled water. The dissolved precipitate was put in a semi permeable cellophane tube, and each end of tube was tied in a distilled water beaker. The distilled water was frequently changed to dialyse out ammonium sulphate ions completely. The contents of the cellophane tube were dissolved in 5 ml of normal saline. The globulin concentration was measured in spectrophotometer and the O.D. values obtained was multiplied by 0.70 (conversion factor to get total immunoglobulin concentration).

2.5.3. Hematology

2.5.3.1 Total erythrocyte count (RBC)

It was done by Hayem's dilution method.

2.5.3.2 Haemoglobulin (Hb) estimation

It was estimated by Sahil's acid haematin method.

2.5.3.3 Total leukocyte count (WBC)

It was done by standard dilution method.

The results were analysed as per Snedecor and Cochran (1994).

3. RESULTS AND DISCUSSION

3.1 Antibody titre

Dexamethasone caused significant decrease in the antibody titre (14.17 ± 2.17) when compared to all other groups. This is in accordance with the findings of Pruett *et al.*, (1987). Levamisole produced significant increase in titre when compared to control and was able to overcome the dexamethasone suppressed antibody titre. Similar results were obtained by Babiuk and Mishra (1981).The control, *C occidentalis* and Dexamethasone + *C occidentalis* groups showed relatively similar values in antibody titre (56.17 ± 0.03, 55.00 ± 7.10, 45.83 ± 6.71 respectively). Dexamethasone + Levamisole groups showed significant increase in antibody titre (183.33 ± 29.02).

Table 1: Effect of immunomodulators on serum antibody titre

Groups	7th day	14th day	21st day	Group
Control	52.54±0.02	76.06±0.06	96.07±0.06	56.17[c]
Dexamethasone	10.00±4.47	20.00±0.00	26.67±4.22	14.17[d]
Levamisole	73.33 ±	133.33±16.87	213.33 ±	105.00[b]
Cassia occidentalis	53.33±8.43	73.33 ±6.67	93.33 ±	55.00[c]
Dexamethasone	120.00 ±	293.33 ±	320.00 ±	183.33[a] ±
Dexamethasone + Cassia	50.00	66.67± 8.43	66.67 ± 8.43	45.83[c] ± 6.71

3.2. Total immunoglobulin concentration

There were no significant differences between dexamethasone, *C occidentalis*, dexamethasone + *C occidentalis* groups (32.39 ± 0.19mg/ml, 32.69 ± 0.13mg/ml, 32.76 ± 0.42 mg/ml respectively). The dexamethasone group showed significant decrease in total serum immunoglobulin level when compared to control group. The inhibitory effect produced by dexamethasone is in accordance with the findings of Anderson *et al.*, (1999). Dexamethasone + Levamisole showed significant increase in the total serum immunoglobulin levels (41.40 ± 0.21 mg/ml). *C occidentalis* caused a decline in immunoglobulin level and it could not restore the immunoglobulin level in immune suppressed animals.

Table 2: Effect of immunomodulators on serum immunoglobulin concentration (mg/ml)

Groups	7th day	14th day	21st day	Group means ± SE
Control	33.01 ± 0.023	33.25 ± 0.02	33.27± 0.004	32.94[c] ± 0.01
Dexamethasone	31.78±0.16	32.92 ± 0.05	32.50 ± 0.28	32.39[cd]± 0.19
Levamisole	35.63 ± 0.19	35.74 ± 0.17	36.56 ± 0.53	35.25[b] ± 0.28
C occidentalis	33.56 ± 0.19	33.08 ±0.04	32.89 ±0.26	32.69[c] ± 0.13
Dexamethasone + Levamisole Levamisole	39.97 ± 0.13	43.63 ± 0.16	48.94 ± 0.09	41.40[a] ± 0.21
Dexamethasone + *C occidentalis*	40.41±= 0.20	44.39 ± 0.16	47.70 ± 0.17	32.76[c] ± 0.42

3.3. Hematology

Dexamethasone produced decrease in WBC levels. This is in accordance with the findings of Wilkie *et al.*, (1979). There were no changes in the RBC and Hb levels between groups. It is obvious that the body haem factors were not stimulated like immune factors by the drugs.

□□□

Table 3: Effect of immunomodulators on haematological parameters

Groups	Total RBC(X10$^{6)}$cmm	Hb (g/dl)	Total WBC(cmm)
Control	5.88±0.02	12.29±0.06	7873.79^{a}±22.65
Dexamethasone	5.82±0.03	12.15±0.10	7354.42^{b}±31.79
Levamisole	5.89 ±0.03	12.32± 0.12	7902.96^{a}±53.15
Cassia occidentalis	5.84±0.07	12.19±0.15	7767.50^{a}±16.99
Dexamethasone + Levamisole	5.90±0.04	12.30± 0.10	8034.67^{a}±49.46
Dexamethasone + Cassia Occidentalis	5.84 ±0.05	12.27±0.15	7728.79^{a}±109.46

4. CONCLUSION

The estimation of antibody titre in the study suggests that the methanolic extract of *C occidentalis* has the potential to improve immunity in rabbits whereas the levels of immunoglobulin could not be reversed by *C occidentalis* in the presence of dexamethasone in rabbits.

REFERENCES

Anderson, B. H., Watson, D. L. and Colditz, I.G., 1999. The effect of dexamethasone on some immunological parameters in cattle. *Vet. Res. Commun* 23(7), 399-413.

Babiuk, L.A. and Misra, V., 1981. Levamisole and bovine immunity: in vitro and in vivo effects on immune responses to herpes virus immunization. *Can.J.Microbiol.* 22, 1312-1319.

Ghani, M.N., 1920. Khazeenatul Advia, Vol. III. Munshi Naval Kishore, Lucknow, p 330.

Jafri, M.A., Jalis subhani, M., J. Kalin., S. Singh., 1999. Hepatoprotective activity of leaves of Cassia occidentalis against paracetamol and ethyl alcohol intoxication in rats. *J. Ethnopharmacol.*, 66, 355-361.

Kabiruddin, M., 1951. Makhzanul advia Shaikh Mohd. Bashir and Sons, Lucknow, pp 454-455.

Pruett, J.H., Fisher,W.F and Deloach,J.R., 1987. Effects of dexamethasone on selected parameters of the bovine immune system. *Vet. Res. Commun.* 11, 305-323.

Snedecor,G.W. and Cochran,W.G., 1994. Statistical methods, 6th Edition, IOWA State Universtiy press, Ames, IOWA.

Talwar. and Gupta, P., 1983. A handbook of practical immunology. Vikas Publishing House Pvt. Ltd., New Delhi.

Tona, L., Ngimbi, N. P., Tsakala, M., Mesia, K., Cimanga, K., Apers, S., De Bruyne, T., Pieters, L., Totte, J., Vlietink, A.J., 1999. Antimalarial activity of 20 crude extracts from nine African clinical plants used in Kinshasa Congo, *J. Ethnopharmacol.*, 68: 193-203.

Wilkie, B.N., Caoili, F. and Jacobs, R. (1979). Bovine lymphocytes Erythrocyte rosettes in normal lymphomatous and corticosteroid-treated cattle. *Can. J. Comp. Med.* 43: 22-28.

27

Pharmacognostic and Physico-chemical Standardization of *Pyrus malus* Linn.

P. Padma Rao, P. Subramanian, P. Sudhakar, T. Sheshashena Reddy and P.R. Reddy

1. INTRODUCTION

Pyrus malus Linn. commonly called as 'apple' in English and 'seb' in Hindi locally, is a well known tree belonging to family Rosaceae. It is indigenous to Europe and West Asia and grow wild in N.W. Himalayas and also cultivated. The leaves and young aerial parts are useful medicinally[1]. Leaves contain phloretin which is a natural antibiotic. Crushed leaves are used as temporary treatment for wounds. In Homoeopathy, its use is mentioned for labyrinthine vertigo by Dr. Cooper, in Boericke[3]. The plant is a tree growing up to 20 to 40 feet, with spreading branches. The leaves are ovate or oblong ovate, serrate, 2-3" long; Petioles ½ to 1" in length. The flowers are large, fragrant, pale rose in colour and borne in sub umbellate corymbs. Fruit is a pome and edible.

Chemically the plant is reported to contain patulin, campesterol, stigmasterol, β- sitosterol, nicotinic acid, phloretin 2- O-glucoside, phlorizin, Gibberellin A 62, melibiose, ostreasterol, 1,28-octacosanediol, pomonic acid and 1,24-tetracosanediol.

The earlier studies on *Pyrus malus* L., pertaining to pharmacognostic and physico-chemical parameters in general and in homoeopathic perspective in particular are not available. Especially, physico- chemical studies in Homoeopathic perspective is not done. Hence, the authors have carried out detailed pharmacognostic and physico- chemical studies on leaves and young aerial parts as per the protocols suggested by Central Council For Research in Homoeopathy (CCRH).

2. MATERIALS AND METHODS

Pharmacognosy

The leaves and young aerial parts of *Pyrus malus* L. was supplied by Survey of Medicinal Plants and Collection Unit of CCRH, Nilgiris, boiled and fixed in F.A.A. (Formaldehyde: Acetic acid: Alocohol) and processed for miocrotomy (Paraffin, method) sectioned, stained and permanent slides prepared following Johansen. The epidermal peels of leaf were obtained by gently scraping and peeling with razor blade. The microtome sections (T.S. and L.S.) were obtained at 6-8 μm thickness on Leica RM 2155 microtome. The powder microscopy characters were studied by boiling the powder drug in distilled water, stained in saffranin and mounted with glycerine. Photomicrography was done with Olympus BX 53 Research trinocular microscope.

Physico-Chemical

The air dried young aerial parts of the drug were coarsely powdered to 10/44 (sieve size) and was subjected to determination of moisture content (loss on drying at 105°C), total ash, water soluble ash, acid insoluble ash, extractability in different solvents, physico-chemical constants, TLC and UV aspects of mother tincture following official methods. Mother tincture was prepared as per HPI. In this method, 100 g of coarse powder of the drug was suspended in 520 ml of 95% alcohol and 500 ml of purified water for 24 hours at room temperature. It was filtered and made up to 1000 ml using same solvent ratio. Percolation method was used for the preparation of mother tincture.

HPTLC analysis

Evaporate 25 ml mother tincture on water bath to remove alcohol. Extract the residue with 3 × 25 ml chloroform. Concentrated chloroform extract is used for the HPTLC study. The concentrated chloroform extract was spotted in the form of band of width 4 mm with a Camag microlitre syringe on precoated silica gel aluminium plate 60F-254, (5 x 10 cm with 0.25 mm thickness; Merck, Darmstadt, Germany) using a Linomat IV sample applicator (Camag, Muttenz, Switzerland, supplied by Anchrom technologists, Mumbai). A constant application rate of 6 ml/sec was followed. The slit dimension was kept at 4 × 0.45 mm and 20 mm/sec scanning speed was employed. The mobile phase consisted of chloroform: methanol (9:1 v/v) and 10 ml of mobile phase was used for chromatography. Linear ascending development was carried out in a 10 x 10 cm twin trough glass chamber (Camag, Muttenz, Switzerland) saturated with the mobile phase at room temperature. The length of the chromatogram run was 8 cm and subsequent to the development, the TLC plates were dried in a current of air with the help of hot air dryer in a wooden chamber with adequate ventilation. Densitometric scanning was performed (Camag TLC scanner III) at 254 nm and 366 nm by reflectance scanning and operated by cats 4 software (v 4.05,Camag) resident in the system.

3. OBSERVATIONS AND RESULTS

Macroscopy: Leaves 2-3" long 1.5-2" wide, ovate or oblong ovate, margins serrate, acute, glabrous above and tomentose beneath, petiole short, 0.5-1" long; young stem greenish, rough and hairy.

Microscopy

Leaf-Surface

Adaxial epidermal cells 5-7 sided, polygonal isodiametric and anisodiametric, sides thin, straight to curved, few sinuate, contents slightly dense, 2687 per sq.mm. Abaxially cells are similar except sides wavy to sinuate; 5560 per sq.mm.

Stomata confined to abaxial , anomocytic, subsidiaries 4 or 5, indistinct, 760 per sq.mm. Stomatal Index 12.02; sizes 22-33 µm long and 11-22 µm wide.

Trichomes are of two types i) uniseriate macroform flagellate conical hair, occur all over and more on veins on abaxial and ii) unicellular filiform cylindrical hair, occur all over on abaxial (Fig.1, 1-3).

TRANSECTION

Midvein is shield like, flat or indistinctly ridged on adaxial and conspicuously ridged on abaxial, 900-1360 µm (1090) in diameter, lamina dorsiventral, 130-270 µm (179) thick, undulated and grooved over smaller bundles (Fig.2.2).

Epidermis is 1- layered, cells adaxially larger on lamina, barrel shaped, tabular, oval to spherical, contents dense in few, abaxially narrower, interspersed with stomata, contents dense. Epidermal cells over midvein, oval to oblong and spherical, papillated and covered by thick cuticle. Collenchyma is hypodermal as a group of cells in midvein adaxial and 1-2 layered on abaxial, lamellar. Ground parenchyma in midvein 15-20 celled in midvein adaxial and 10-12 layered on abaxial, cells polygonal to spherical 8-32 µm in dia. on adaxial and 22-63 µm in dia. on abaxial. Palisade 2- layered, closely packed, cylindrical cells extending into midvein. Palisade up to 7-11. Spongy tissue nearly half the area of lamina with oval to oblong, cylindrical and dumbbell shaped loosely dispersed cells (Fig.2.1).

Vascular tissue of midvein consists of single large arc shaped vascular bundle, 680-790 ìm (730) long and 420-560 ìm (490) wide, endarch, conjoint, collateral with a cambium between xylem and phloem. Xylem is extensive, arranged in radial rows. Tracheary elements in L.S. are mostly helical and few scalariform thickened. Phloem is external and precocious and surrounded by discontinuous patches of sclerenchyma (Fig.2.2).

Petiole

In T.S. shield like with shallow groove on adaxial and two slender ridges on either sides; 792-1330 μm (1129) long and 736- 1189 μm (988) wide.

Epidermis is 1- layered, cells small, oval to spherical, papillated and covered by thick cuticle. Hypodermis is 1-2 layered, collenchymatous, angular or lamellar thickened. The peripheral cortex is aerenchymatous, 8-15 layered, cells polygonal to spherical, 11-49 μm in diameter. Contents scanty, in few dense (Fig.2.3).

The vascular tissue is made of a large arcuate bundle at centre besides two small spherical adaxial bundles.(Fig.2.3)

The central v. bundle is laterally 529-842 μm (678) long and 216-432 μm (313) wide, endarch, conjoint, collateral with a cambium in between xylem and phloem. The vascular bundle is enclosed by thick layer of sclerenchyma. The adaxial bundles are 82-48 μm (128) in diameter and enclosed by scerenchymatous sheath (Fig.2.3).

Young Stem

In T.S. stem is oval to spherical, 1.8-2 mm in diameter. The outermost epidermis is 1- layered. Collenchyma is 1-3 layered, cells oval to spherical and tabular, lamellar or angularly thickened. Cortex is 12-15 layered with peripheral cells containing chloroplasts. The vascular tissue is in the form of continuous cylinder. The xylem is in transition state with secondary growth in some v. bundles. Vessels/ tracheids mostly isolated and few in radial rows of 2-3. Medullary rays are 1-3 seriate. The phloem is external and interspersed with prismatic crystals. Vascular cambium is also present. Discontinuous patches of sclerenchyama encloses the phloem. Secretory canals are arranged in a ring. Pith is abundant at the centre (Fig.3,1-4).

Powder Microscopy

1. Fragments of leaf with unicellular conical and cylindrical hair.
2. Flagellate hairs, many.
3. Crystals of calcium oxalate, few.
4. Fragments of epidermis of lower surface with anomocytic stomata.
5. Pieces of tracheary tissue, with fibers, vessels and attached ray parenchyma.
6. Fragments of cortical parenchyma with dense contents.

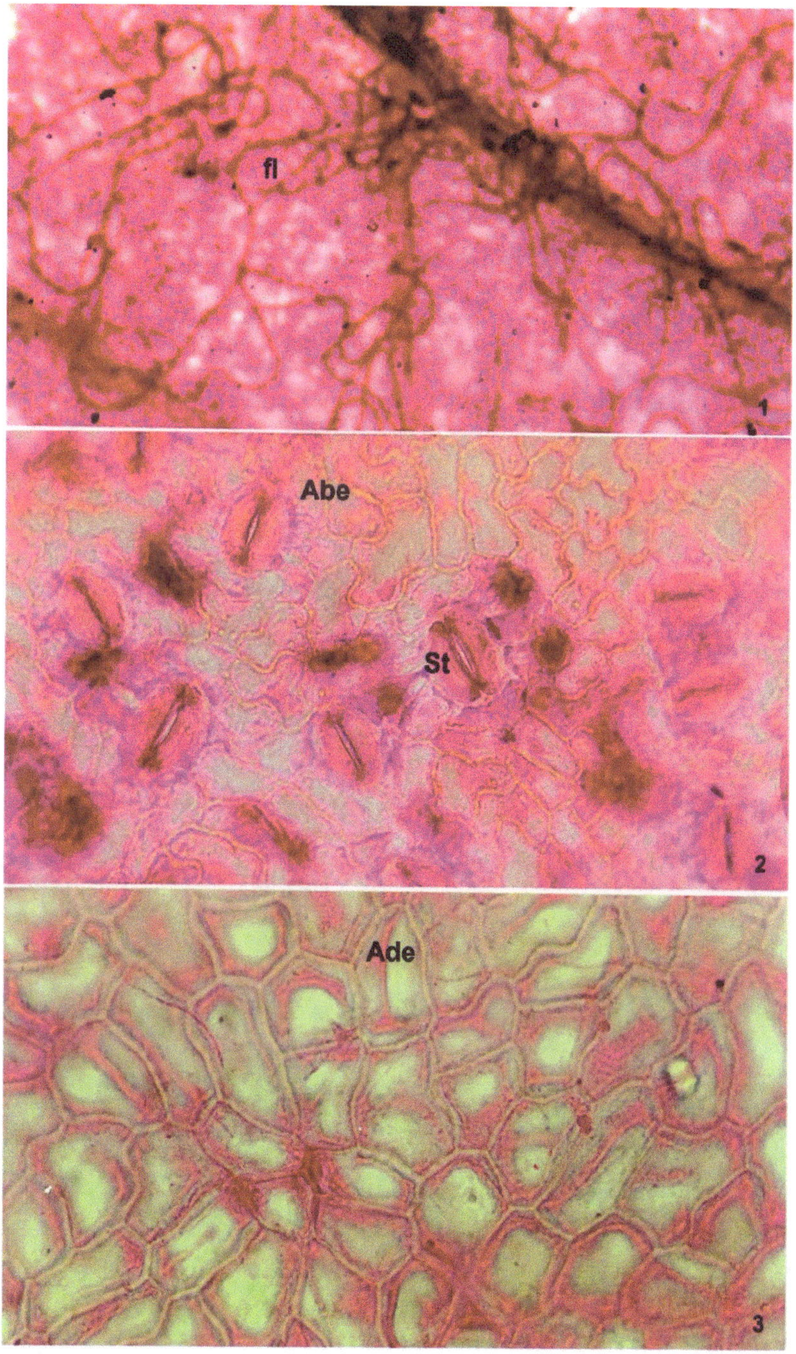

Fig. 1: *Pyrus malus* (leaf)

(1) Epidermis in surface with flagellated hair x 47 **(2)** Epidermis of abaxial surface x 600 **(3)** Epidermis of adaxial surface x 634

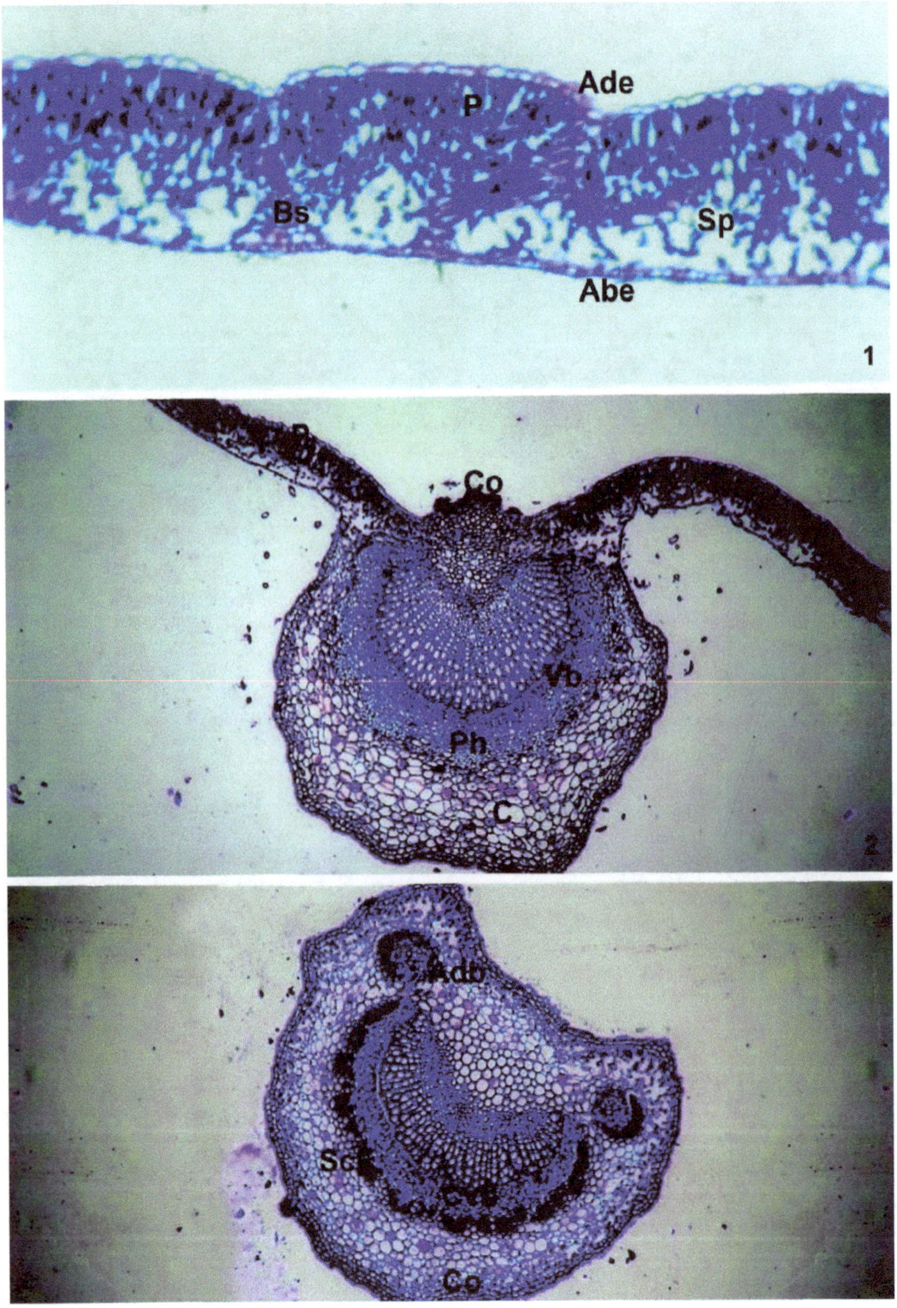

Fig. 2: *Pyrus malus* (leaf)
(1) T.S. of leaf lamina x 138 **(2)** T.S. of leaf at midvein x 43 **(3)** T.S. of petiole x 55.5

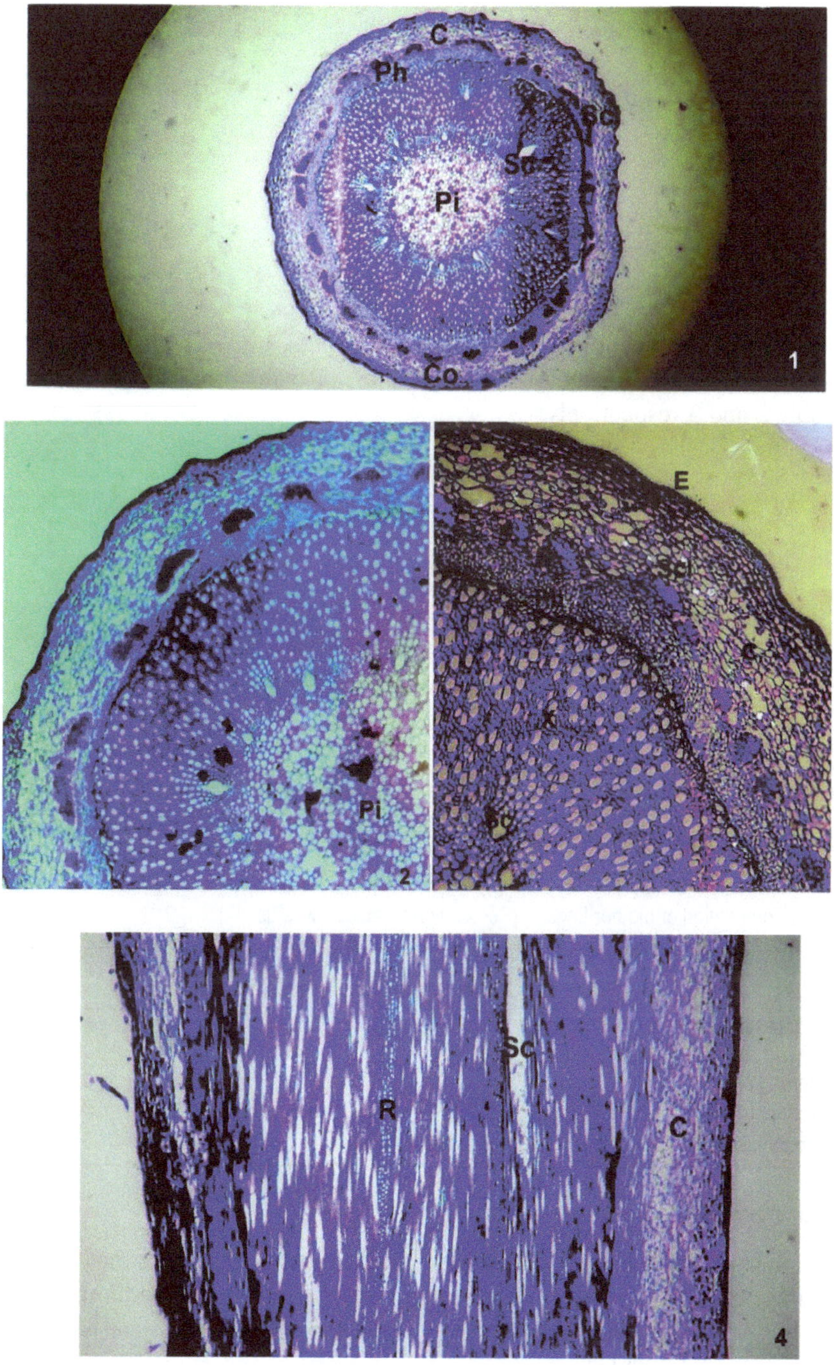

Fig. 3: *Pyrus malus* (young stem)

(1) T.S. of young stem x 25 **(2)** T.S. of young stem, a portion x 56 **(3)** T.S. of young stem, a portion x 43 **(4)** L.S. of young stem x 43

ORGANOLEPTIC CHARACTERS

Colour - Brownish yellow.

Touch - Smooth.

Odour - Pungent.

Taste - Bitter.

PHYSICO-CHEMICAL STUDIES

The determined data under the physico-chemical study for the raw drug is summarized in Table 1 and that of mother tincture preparation and its standardisation in Table 2 and 3 respectively.

Table 1: Standardisation of Raw Drug

S.No	Parameters	Quantitative values
1.	Moisture content (Loss on drying at 105° C)	Not more than 6.34% w/w
2.	Total ash	Not more than 3.6% w/w
3.	Acid insoluble ash	Not more than 0.43% w/w
4.	Water soluble ash	Not more than 2.45% w/w
5.	Alcohol soluble extractive	Not less than 17.0% w/w
6.	Water soluble extractive	Not less than 17.9% w/w
7.	Extractive values in	
	a. Hexane	Not less than 0.93% w/w
	b. Chloroform	Not less than 2.15% w/w
	c. Methanol	Not less than 21.5 % w/w

Table 2: Formulation of mother tincture (Percolation technique was used[8].)

Drug strength	1/10
Preparation:	
Pyrus malus in coarse powder	100 g
Strong alcohol	520 ml
Purified water	500 ml
To make one thousand millilitres of the mother tincture	

Qualitative phytochemical tests:

Preliminary phytochemical study indicates the presence of alkaloids, steroids, saponins, and tannins. Moisture content reveals the presence of water in the plant and also some volatile organic matter. Results of physico-chemical studies are summarized in Tables1-3.

Table 3: Standardisation of Mother Tincture

S.No.	Parameters	Observations
1.	Organoleptic profile	
	a. Appearance	Clear, non viscous, foam on shaking
	b. Colour	Dark brown
	c. Odour	Pleasant and fruity
2.	Sediments	Absent
3.	Weight per ml	Not more than 0.93 g
4.	Total solids	Not less than 1.26 % w/v
5.	Alcohol content	47-50 % v/v
6.	pH	4.5-5.0
7.	ë max	223, 287, 327 nm

HPTLC finger printing

The profile of chromatographic separation scanned at 254 nm shown in Figure, reveals ten spots. Among them 10, 9 and 8 spots possess maximum composition with R_f at 0.80, 0.76 and 0.68 respectively. It is clear that these are characteristic for the studied drug, which will help in identification and authentication of the mother tincture. This is considered as valuable standards in pharmacopoeia. While, chromatogram scanned at 366 nm reveal ten spots with 3, 4 and 10 spots showing maximum composition at R_f0.35, 0.42 and 0.88 respectively. At 254 nm, 10 spots appear at R_f 0.10, 0.20, 0.28, 0.36, 0.42, 0.52, 0.59, 0.68, 0.76 and 0.80 (Fig.5) with various concentrations while at 366 nm, ten spots at R_f 0.11, 0.23, 0.35, 0.42, 0.54, 0.59, 0.66 , 0.80, 0.84 and 0.88 (Fig. 4).

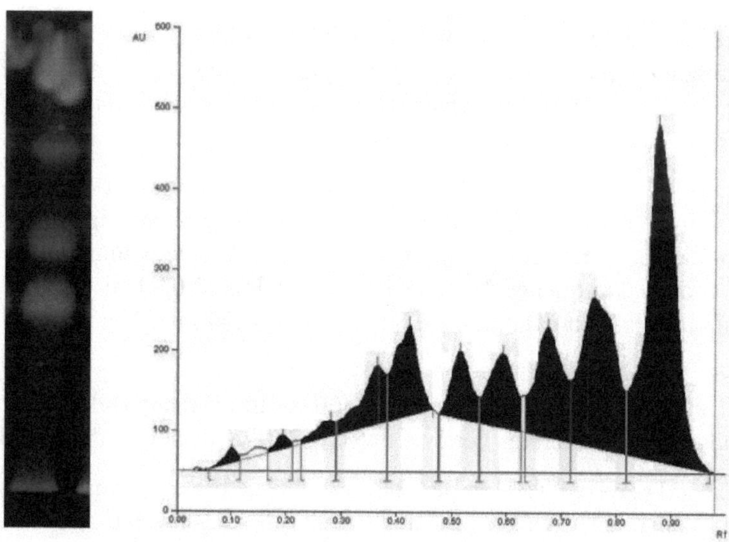

Fig 4: HPTLC finger printing of *Pyrus malus* mother tincture chloroform extract scanned at 366 nm.

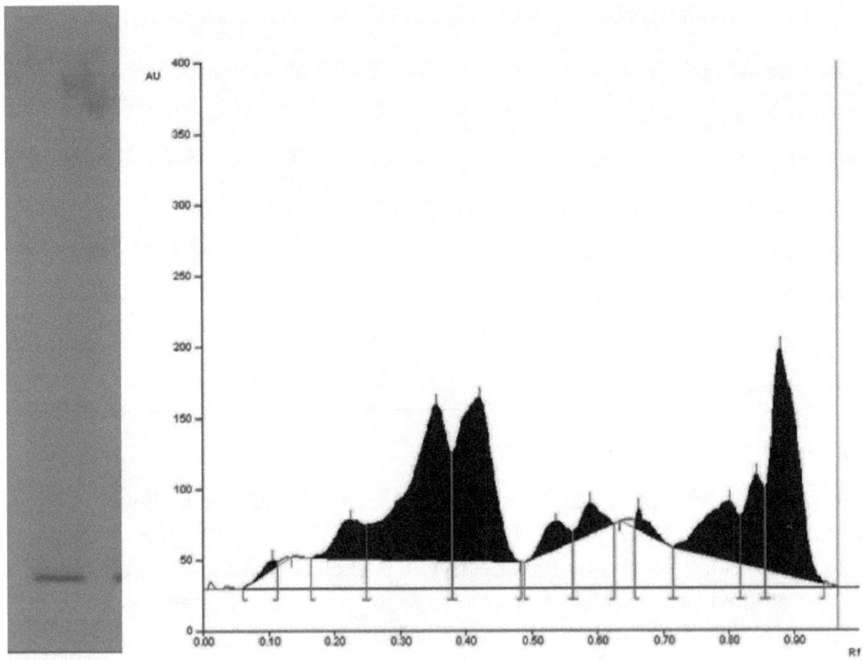

Fig 5: HPTLC finger printing of *Pyrus malus* mother tincture chloroform extract scanned at 254 nm.

4. DISCUSSION

Pyrus malus Linn. popularly known as 'apple' is a tree growing up to 20-40 ft. high belonging to the family Rosaceae. Morphologically the leaves are 2-3" long ovate or oblong ovate, serrate, glabrous above and tomentose beneath. Petiole is up to 1" long and young stem is greenish, rough and hairy.

Epidermal cells in adaxial surface possess sides thin, straight to curved, few wavy to sinuate walls and 2687 per sq.mm. while those on abaxial have wavy to sinuate walls and 5560 per sq.mm. Leaf is hypostomatic with anomocytic type of stomata. Stomatal frequency 760 per sq.mm and S.I. is 12.02. Uniseriate flagellate conical and unicellular cylindrical hair occur on abaxial side. The palisade ratio is 7-11.5.

The midvein in T.S. is shield like, 900-1360 ìm in diameter and conspicuously ridged on abaxial. Lamina is dorsiventral 130-270 ìm thick and undulated.

Epidermis is 1- layered with larger cells on adaxial. The epidermal cells over the midvein are papillated with thick cuticle. Palisade is 2- layered and extend into the midvein. Spongy tissue occupy half the mesophyll and possess dispersed cells.

A single large arcuate bundle, endarch, collateral with a vascular cambium is present in the midvein. A layer of discontinuous patches of sclerenchyma is present enclosing the v. bundle.

Petiole

Shield like with a shallowed groove on adaxial. Centrally a large acruate bundle and with two small spherical bundles is present in adaxial ridges. Peripheral cortex is aerenchymatous. A sclerenchymatous sheath encloses the V. bundles.

Young Stem

Oval to spherical in out line and up to 2 mm in diameter. The cortex is 12-15 layered. The vascular tissue is in the form of a continuous cylinder. The xylem also shows secondary growth in the v. bundles. Medullary rays are 1-3 seriate. The phloem is external and interspersed with prismatic crystals of calcium oxalate. Secretory canals are present towards centre. Pith is abundant.

The powder microscopic features and organoleptic characters along with the anatomical studies are diagnostic and establish the standards for the drug.

Physico- chemical

The values of total ash, water soluble ash and acid insoluble ash are falls within acceptable range. The results of HPTLC finger printing and UV Spectrophotomatric study exhibits (Maximum absorption) three prominent peaks which will be taken as characteristic standards for the drug.

5. CONCLUSION

The determined physico-chemical data, macro and microscopical characters and methodology employed in the study will help in identification, authenticity and ensures quality, purity and efficacy of the drug.

ACKNOWLEDGEMENTS

Dr. R.K. Manchanda, Director General, Central Council for Research in Homoeopathy. New Delhi, for the facilities and encouragement; Dr. Uday Kumar, Deputy Director, Pathology Division, National Institute of Nutrition, Hyderabad for the microtome facility; and Director, Indian Institute of Chemical Technology, Hyderabad for the HPTLC facility.

REFERENCES

Anon. 1971. Homoeopathic Pharmacopoeia of India, Vol. I, Min. of Health & F.W., Govt. of India.

Anon., Wealth of India, Raw materials, Publications and Information Directorate, CSIR., New Delhi. 1956.

Boericke, W. Pocket Homoeopathic Materia Medica and Repertory (Repr. Ed) Jain Publishing Co., New Delhi, p.p.465.

Christian Gosch, Heidi Halbwirth and Karl Stich, Phytochemistry, 2010, 71(8-9), 838-843.

Johansen, D.A., Plant Microtechnique, Macgraw Hill. Book Co. New York, 1940.

Kirtikar, K.R. and D. Basu, Indian Medicinal Plants. Bishen Singh Mahendrapal Singh. Dehradun., Vol.2.,p.p.987,1980.

Nadkarni.K.M., Indian Materia Medica (Rev.& Enlarg.) Bombay Popular Prakashan, Bombay, Vol. II, p.p. 1039.

Rastogi, R.P. and Mehrotra, B.N. 1960 – 1989. Compendium of Indian Medicinal Plants Vol. I – IV). CDRI., Publication and Information Directorate, New Delhi

Sethi. P.D., High Performance Thin Layer Chromatography-CBS Publishers and Distributers, NewDelhi, 1996.

Stahl.E.1969. Thin Layer Chromatography, A laboratory hand book, Springer- Verlag, Berlin.

Wagner.H *et al.* 1984. Plant Drug Analysis, A TLC Atlas, Springer- Verlag, Berlin.

28

Studies on the Anti-inflammatory and Analgesic Activity of Selected Seaweeds Macro Algae in Thondi Coast

M. Dhinakarraj, S. Ravikumar, M. Anand and M. Babuselvam

1. INTRODUCTION

Inflammation have been linked with pathogenesis of many diseases like Cancer, Atheriosclerosis, Neurodegenerative diseases, Diabetes Mellitus, Obesity, Arthritis, Cardiovascular diseases, Parkinson's disease and other deadly diseases (Lee and Weinblatt, 2001). Inflammation can result to genetic defects and immunoregulation and mechanism defects which lead to tissue damage (Jones, 2006). Anti-inflammatory drug was serendipity discovered without thorough investigation and even not effective in term of controlling inflammation and now record several side effect just like other synthetic drugs, thus, safe biological sources are now been considered. Biological sources for active compounds that have medical importance are increase in recent time (Kaboli, 2001). Also, due to the residual side effects of synthetic compounds that form bulk of the materials used for the production of pharmaceutical products. Aspirin can cause stomach bleeding, acetaminophen can cause liver damage Cox-2 inhbitor Vioxx and Celebrex can cause heart problem and non-steroidal anti-inflammatory drugs (NSAID's) was reported to contribute to numerous death (Clegg *et al.,* 2006; Wolf *et al.,* 1999; Sing, 1998).

Plants have been a rich source of medicines because they produce a variety of bioactive molecules, most of which are probably evolved as chemical defences against predation (or) infection. Medicinal plants contain active constituents like terpenes, alkaloids, steroidal saponins, tannins, phenols, quinones and flavonoids (Levin and York, 1978; Reinhold *et al.,* 1981; Horborne, 1982; Bever, 1986). The potential use of higher plants as a source of new drugs is still poorly explored. Of the estimated 250,000-500,000 plant species, only a small percentage has been investigated phytochemically (Payne, 1991). Ethnobotanical and ethnopharmacological research are crucial in the development of drugs from natural sources (Goldfrank, 1982; Vulto *et al.,* 1988; Mentz and Schenkel, 1989). The information obtained on identification, preparation, clinical use, gathering and preservation of medicinal plants dramatically facilitate the search for new drugs, which are needed for the drug development programs (Farnsworth and Soejarto, 1985; Soejarto and Farnsworth, 1989). Seaweeds are marine macro algae and primitive type of plants, growing abundantly in the shallow waters of sea, estuaries and backwaters. They flourish wherever rocky, coral, or suitable substrata are available for their attachment. They belong to three groups namely green, brown and red based on pigmentation, morphological and anatomical characters. Selective utilization of marine algae as potential source of pharmaceutical agents has been increasing in recent years. Many of the seaweeds possess bio-active components which inhibit the growth of some of the Gram positive and Gram negative bacterial pathogens. The algal extracts were used as a curative and preventive agent for various diseases such as AIDS, Diabetic, Tumor, Dengue and Malaria.

Most of the bioactive substances isolated from marine algae are chemically classified as brominated, aromatics, nitrogen-heterocyclic, nitrosulphuric-heterocyclic, sterols, dibutanoids, proteins, peptides and sulphated polysaccharides. The crude extract thus obtained is subjected to broad based biological screening for antifungal, antiviral, antibacterial, antimalarial, antifilarial, hypoglycaemic and antifertility activity and anti-inflammatory activity (Garg 1993; Ravikumar *et al.,* 2002;2012). So, the present investigation was aimed to asses the therapeutic potential of marine halophytes of seaweeds viz, *Gracilaria edulis, Sargassum wightii* and *Ulva fasciata* for the development of alternative therapeutics for the treatment of inflammatory, analgesic and arthritic diseases.

2. MATERIALS AND METHODS

2.1 Antiinflammatory activity of Seaweed marine halophytes

2.1.1 Collection and extraction of bioactive compounds from chosen seaweeds

Fresh samples of *Gracilaria edulis, Sargassum wightii* and *Ulva fasciata* were collected from Thondi coastal area (latitude of 9° 44' N and longitude of 79° 02' E). Samples were washed thrice in sterile distilled water to remove adhering soil particles and salts. All the samples were shade dried and were powdered. 500 g of powdered leaves were soaked with 1 litre of ethanol water mixture (3:1) for 10 days in an air tight clean glass container (percolation). Mixing was done every day so as to enable to enhance the maximum extraction of bioactive compounds. After that, solvent with bioactive compounds were filtered with Whatman filter paper. The filtrate was then subjected to the rotary flash evaporator (BUCHI, Japan) and further lyophilized (BENCHTOP 2K) to remove excess organic residue. The residual extracts were further used for the anti-inflammatory analgesic and anti-arthritic activity. After filtration, the suspended materials were dried under room temperature and weighed. The percentage of extraction was calculated as follows:

$$\text{Percentage of extraction (\%)} = \frac{\text{Weight of the extract (g)}}{\text{Weight of the plant material (g)}} \times 100$$

2.1.2 Carageenan method

The anti-inflammatory activity was examined in rats according to the method of Winter *et al.,* (1962) with slight modifications due approval of animal ethical clearance was obtained from Madurai Kamaraj University, Madurai. The rats were randomly divided into 6 groups of 6 animals. Before treatments, the average volume of the right hind paw of each animal was measured 3-4 X by plethysmometer 7150 (Ugo Basile). These measurements did not differ by more than 2% and represent the initial paw volume. The compounds i.e. vehicle (0.9% saline, 10% DMSO as 0.5 ml $100g^{-1}$ bw), standard drug (aspirin in 100 mg kg^{-1} bw) and the most effective seaweed marine halophytic extracts *viz., Ulva fasciata, Gracilaria edulis* and *Sargassum wightii* plant extracts were administered orally to the animals. Thirty minutes after the administrations of test compounds, each rat received in its right hind paw a sub planter injection of a 1% carrageen suspension (0.05ml per animal) through 26- gauge needle. The average volume of the right hind paw of each rat was determined as described above at 1st, 2nd, 3rd and 4th h after the injections of carrageenan. The oedema volume and percentage of inhibition in oedema were calculated as described below.

Inflammation of paw volume in rats was calculated by taking the difference between the final (Vt) i.e. after carrageenan and initial (Vo) i.e. before carrageenan administration and average volumes (Vt-Vo) for each group at different time intervals. Anti-inflammatory effects were calculated as percent inhibition in oedema, obtained at each time intervals for each group, using the following formula (Lanhers *et al.,* 1992).

$$\text{\% of inhibition of oedema} = \frac{[(\text{Vt-Vo})\ \text{control} - (\text{Vt-Vo})\ \text{treated}]}{(\text{Vt-Vo})\ \text{control}} \times 100$$

The data was represented as mean ± standard error of mean (SEM) or as percentages. Results were statistically evaluated using analysis of variance (ANOVA) involving effects of different factors like nature of extracts on the anti-inflammatory activities.

2.1.3 Cotton pellets – induced granuloma

The rats were divided into five groups (n=6). The cotton pellet granuloma model investigated the proliferation phase of inflammation (Vorkoc and Sedmera, 1977). The extract of selected seaweed extracts at different doses (250 and 500 mg kg^{-1} bw) and aspirin 100 mg kg^{-1} bw was given to the animals orally. After 30 minutes the animals were anesthetized and 10 mg of sterile cotton pellets were inserted, one in each axilla. The extracts were administered daily for a period of seven days. The rats were sacrificed after a high dose of anesthesia on the eighth day and the cotton pellets were removed surgically and made free from extraneous tissues. The pellets were incubated at 37°C for 24 h and dried at 60°C to constant weight (Winter and Poster, 1957). Increment in the dry weight of the pellets was taken as a measure of granuloma formation.

2.2 Evaluation of analgesic activity of seaweed extracts

2.2.1 Acetic acid writhing method

The abdominal constriction test described by Collier *et al.,* (1968) was used to measure the antinociceptive actions of test extracts and aspirin. Test animal weighing between 200–250 g were fasted for 24 h with water given *ad-libitum* and were pretreated with oral 4% acacia gum solutions (0.15 ml $10g^{-1}$), aspirin (100 mg kg^{-1} bw) and tested seaweed extracts were given 30 minutes prior to intraperitonial injection of 0.15 ml $10g^{-1}$ of 0.6% acetic acid to cause typical stretching response. Writhing or stretching (abdominal constrictions) was counted for a period of 5 minutes under double blind observations. The antinociceptive effects of drugs were measured by calculating the mean reduction in the number of abdominal constrictions for each drug, as compared to gum acacia controls.

2.2.2 Hot plate method

Test animal weighing between 200-250 g was fasted for 24 h with water given *ad-libitum*, maintained at room temperature and were divided into 4 groups of six rats. Rats were treated with 4% acacia gum (0.15 ml $10g^{-1}$), aspirin (100 mg kg^{-1} bw) and seaweed extracts of different concentrations of 250 and 500 mg kg^{-1} bw. Analgesic effect of the test extracts were determined by the hot tail flick method described by Swell and Spencer (1976). One to two cm of the tail of rat was immersed in warm water kept constant at 50°C. The reaction time was the time taken as a mean of the next two readings. The latent period of the tail-flick response was taken as the index of antinociception and was determined before and at 15,

30, 45 and 60 min after the administration of drugs. The maximum reaction time was fixed at 15 seconds. The Maximum Possible Analgesia (MPA) was calculated as:

$$MPA = \frac{\text{Test reaction time} - \text{Saline reaction time}}{15 - \text{Saline reaction time}}$$

3. RESULTS

3.1 Anti-inflammatory activity of seaweed extracts by carrageenan induced paw edema in rats

The present study observed that, control rats (n=6) treated with intraperitoneal administration of 0.9% saline showed a progressive time dependent swelling of the right hind paw after the sub planter administration of 0.05 ml of 1% carrageenan reaching its maximum at 2nd hour. Acute anti-inflammatory studies showed that seaweed extract of two concentration of *Ulva fasciata, Gracilaria edulis,* and *Sargassum wightii* produced a significant inhibition of carrageenan induced rat paw oedema at 4th hour ($p<0.005$) causing 63.15%, 62.15% and 69.69% respectively as compared to the control groups (rats). Intraperitoneal administration of aspirin (100 mg kg^{-1} bw) caused dose dependent inhibition of oedema (Table 1). Among the seaweed crude extracts, *Sargassum wightii* (250 and 500 mg kg^{-1} bw), exhibited significant ($P<0.01$) reductions in paw oedema than the other seaweeds.

In the granuloma pouch test, a popular test of anti-inflammatory agents, an aseptic inflammation with large volumes of hemorrhage is induced. The present study revealed that, in the chronic model (cotton pellet induced granuloma) on the seaweed extracts of *Ulva fasciata, Gracilaria edulis* and *Sargassum wightii* at the dose level of 250 and 500 mg kg^{-1} bw and standard drug aspirin (100 mg kg^{-1} bw) showed decreased granuloma tissue and increased percentage of inhibition. Among the seaweeds, the extract of *Sargassum wightii* at a concentration of 500 mg kg^{-1} bw showed maximum inhibition by 57.03% as compared to control (Table 2).

3.3 Analgesic activity of seaweed extracts by acetic acid writhing model

Analgesic effects of chosen seaweeds *viz., Ulva fasciata, Gracilaria edulis,* and *Sargassum wightii* at different doses of 250 and 500 mg kg^{-1} bw on the writhing test in rats is shown in Table 3. Extract of *Sargassum wightii* at the dose of 500 mg and aspirin at the dose of 100 mg exhibited significant inhibition compared with the control writhes by 84.6 and 66 % respectively.

Table 1: Anti-inflammatory effect of seaweed extracts in carageeenan induced paw edema in rat

Treatments	Dose (mg kg^{-1} bw)	1 hour Increase in paw volume (Mean ± SE)	% of inhibition	2 hour Increase in paw volume (Mean ± SE)	% of inhibition	3 hour Increase in paw volume (Mean ± SE)	% of inhibition	4 hour Increase in paw volume (Mean ± SE)	% of inhibition
Ulva fasciata	250	0.24 ± 0.01	32.43	0.29 ± 0.01	35.53	0.18± 0.01	55.26	0.16± 0.01	57.89
	500	0.22 ± 0.01	43.24	0.17± 0.01	57.50	0.16± 0.01	57.89	0.14± 0.01	63.15
Gracilaria edulis	250	0.23 ± 0.01	35.13	0.21± 0.01	50.00	0.18± 0.01	52.63	0.16± 0.01	57.89
	500	0.19 ± 0.01	48.6	0.16± 0.01	60.00	0.15± 0.01	60.52	0.15± 0.01	62.15
Sargassum wightii	250	0.20 ± 0.01	39.39	0.23 ± 0.01	42.50	0.16 ± 0.01	55.16	0.13 ± 0.01	60.61
	500	0.18 ± 0.01	54.05	0.16 ± 0.01	60.52	0.14 ± 0.01	60.61	0.11 ± 0.01	69.69
Control	-	0.37 ± 0.01	-	0.41 ± 0.01	-	0.39 ± 0.01	-	0.38 ± 0.01	-
Aspirin	100	0.21± 0.01	33.6	0.18± 0.01	52.8	0.17± 0.01	59.8	0.15± 0.01	61.7

Values were found insignificant between treatments and found significant between duration of treatment (P>0.01)3.2 Anti-inflammatory activity of seaweed extracts by cotton pellet induced granuloma in rats

Table 2: Effect of the seaweed extracts on cotton pellets induced granuloma in rats

Treatments	Dose ($mg\ kg^{-1}$ bw)	Dry weight of cotton pellet (mg)	Percentage of inhibition (%)
Control	-	46.60 ± 1.16	-
Aspirin	100	22.4 ± 1.73	56.23
Ulva fasciata	250	28.37 ± 0.87	40.02
	500	24.40 ± 1.15	50.41
Gracilaria edulis	250	23.4 ± 0.85	52.32
	500	21.6 ± 0.81	56.32
Sargassum wightii	250	27.6 ± 1.25	41.13
	500	20.6 ± 1.10	57.03

Values are found insignificant between treatments; ± SD

Table 3: Analgesic activity (acetic acid writhing test) of seaweed extracts in rats

Treatments	Acetic acid writhing in test animal		
	Dose ($mg\ kg^{-1}$ bw)	Number of writhes in 20 minutes	Percentage of inhibition
Control	3 $ml.kg^{-1}$	46.78 ± 1.09	-
Aspirin	100	16.77 ± 1.07	66.00
Ulva fasciata	250	20.49 ± 0.85	51.56
	500	13.63 ± 0.62	72.00
Gracilaria edulis	250	20.50 ± 0.13	51.56
	500	11.05 ± 0.25	74.30
Sargassum wightii	250	19.62 ± 0.87	57.00
	500	7.39 ± 0.88	84.6

Values are found insignificant between treatments; +SD

3.4 Analgesic activity of seaweed extracts by hot plate method

In the hot plate test, an analgesic effect of *Ulva fasciata, Gracilaria edulis,* and *Sargassum wightii* were observed at the concentration of 250 and 500 $mg\ kg^{-1}$ bw. This result indicates that, the seaweed extract possesses both inflammatory and

Table 4: Analgesic effect (Hot plate model) of seaweed extracts in rats

Treatments	Hot plate method		
	Dose ($mg\ kg^{-1}$ bw)	Mean reaction time (S)	Percentage of inhibition (%)
Control	3 $ml.kg^{-1}$	2.646 ± 0.47	—
Aspirin	100	16.57 ± 0.54	62.72
Ulva fasciata	250	4.689 ± 0.52	48.0
	500	6.701 ± 0.39	65.8
Gracilaria edulis	250	3.927 ± 0.24	40.1
	500	5.92 ± 0.24	61.25
Sargassum wightii	250	18.16 ± 0.53	72.72
	500	18.57 ± 0.90	80.29

Values are found significant between treatments ($P>0.05$); ±SD non-inflammatory pain relif. It is proved by the present study that, extract of *Sargassum wightii* at a dose of 500 mg kg^{-1} bw possesses an maximum activity (80.29%) when compared to control (Table 4).

4. DISCUSSION

There are several tissue factors that are known to be involved in the inflammatory action such as release of histamines and prostaglandins. However, development of non-steroidal anti-inflammatory agents in recent years have contributed a lot not only in human sufferings but also helped in understanding the tissue mechanisms of inflammation. The inflammatory reaction is readily produced in rats in the form of paw oedema with the help of irritants or inflamagens. Carrageenan induced paw oedema is the most commonly used method in experimental pharmacology causing inflammation by producing histamine, 5-HT and prostaglandins. Therefore, in the present study, the anti-inflammatory property of the ethanol extract of marine halophytes of seaweeds *viz., Gracilaria edulis, Sargassum wightii* and *Ulva fasciata* were evaluated by challenging the animal with carageenan. Present results showed that, crude extract of *Sargassum wightii* exhibited significant ($p<0.01$) reduction in paw edema in rats.

The factor responsible for such anti-inflammatory property of the extract might be due to the presence of organic substances/compounds like tannin which is capable of inhibiting the flamogen like histamine and prostaglandin system (Muruganandan *et al.*, 2001). Morever, the non-steroid compound like phenylbutazone, is also capable of inhibiting the migration of leucocytes in carrageenan induced pleuracy. The present studies found that the crude extract from *Sargassum wightii* as an inflammatory agent. This is coincides with the inhibition of the production of prostaglandins. This inhibition of synthesis of prostaglandins might be due to the presence of alkaloids, quinines, saponins, protein, steroids and sugars.

Previous studies reported that, plants are rich in alkaloids triterpenoids, flavanoids and saponins which showed potential anti-inflammatory activity (Haihua Shu *et al.,* 2006; Theophile *et al.,* 2006; Borgi, 2007). Preliminary phytochemical analysis performed in this study shows the presence of alkaloids, quinines, saponins, protein, steroids and sugars in the extract of *Sargassum wightii.* The anti-inflammatory activity of triterpenoids have been reported by many researchers (Vazquez *et al.,* 1996; Suh *et al.,* 1998; Huss *et al*., 2002). Moreover, the suppression of inducible nitric oxide synthase (iNOS) and inducible cycloxygenase – α (COX-2) enzymes are due to triterpenoids. Saponnins has also been reported to have anti-inflammatory activities by inhibition of the enzymes iNOS, COX-2 and lipoxygenase (Bermejo Benito *et al.,* 1998; Li and Chu, 1999: Kim *et al.,* 2002). Therefore, it seems that analgesic and anti-inflmmatory profile of *Sargassum wightii* might be due to the presence of saponins in the crude extract.

The analgesic activities of marine halophytes of selected seaweeds were also studied by the present study. It reveals that, extract of *Sargassum wightii* significantly decreased the number of writhes in 20 minutes and also increased the percentage of inhibition in acetic acid writhing test in test animals. Therefore, it is likely that *Sargassum wightii* extracts might suppress the formation of prostaglandins or antagonizes the action of these substances and thus exerts its analgesic activity in acetic acid induced writhing test.

The hot plate method is considered to be selective for opioid like compounds in several animal species (Janssen *et al.,* 1963). It could be suggested that, the extract may possess both anti-inflammatory and analgesic activities probably mediated through common mechanisms. Experimental evidence obtained in the present study indicates that, extract of *Sargassum wightii* (500 mg kg^{-1} bw) significantly delayed the reaction times of the rat used in the hot plate analgesic test method. Moreover, these observations tend to suggest that extract of *Sargassum wightii* possesses centrally and peripherally mediated analgesic properties. The peripheral analgesic effect of the plant extract may be mediated via inhibition of cyclooxygenase and lipoxygenase, while the central analgesic action of the plant extracts may be mediated through inhibition of central pain receptors. Although the results of the present study are inconclusive, they tend to suggest that *Sargassum wightii* extracts are probable produces its anti-inflammatory effect by inhibiting the release, synthesis and/or production of inflammatory mediators, including polypeptide kinins, prostaglandins and so forth, like aspirin. The present investigation suspected that, presence of alkaloids and flavanoids in *Sargassum wightii* posses analgesic effect with less toxicity.

4. CONCLUSION

In conclusion, the results of present study showed that the extract of *Gracilaria edulis, Ulva fasciata* and *Sargassum wightii* possesses anti-inflammatory activity an acute inflammatory processes. Its anti-inflammatory effect could be produced by the alkaloids, flavanoids, saponins and tannins were present in the chosen extracts. The present study showed that *Sargassum wightii* extract also possesses analgesic and arthritic effects, which can be very useful in painful haemorrhoid. The present findings proved the tradition use of *Sargassum wightii* is an anti-inflammatory and analgesic drug in Indian folk medicine. In conclusion, extract of *Sargassum wightii* can be considered as safe to be used as an alternative therapeutics for the treatment of anti-inflammatory and analgesic diseases after completing successful clinical trials with human model.

REFERENCES

Bermejo Benito P, Abad Martinez M J, Silvan Sen A M, Sanz Gomez A, Fernandez Matellano L, Sanchez Contreras S and Diaz Lanza A M, (1998). *In-vivo* and *in-vitro* anti-inflammatory activity of saikosaponins. *Life Sci,* 63: 1147–1156.

Bever B, (1986). Medicinal plants in tropical West Africa. Cambridge University Press, Cambridge, MA.

Borgi W, Ghedira K and Chouchane N, (2007). Anti-inflammatory and analgesic activities of *Zizyphus* lotus root barks. *Fitoterapia,* 78: 16–19.

Clegg D, Reda D, Harris C, Klein M, O'Dell J, Hooper M, Bradley J, Bingham C, Weisman M, Jackson C, Lane N, Cush J, Moreland L, Schumaker H, Oddis C, Wolfe F, Molitor J, Yokum D, Schnitzer T, Furst D, Sawitzke A, Shi H, Brandt K, Moskowitz R, Williams H,(2006). Glucosamine, Chondroitin Sulfate and the two in combination for painful knee osteoarthritis. *New Engl. J. Med.*, 354(8): 795-808.

Collier H O J, Dinneen L C , Johnson C A and Schneider C, (1968). The abdominal constriction response and its suppression by analgesic drugs in mouse. *Braz J Pharamcol and Chemother.*, 32: 295.

Farnsworth, N R., and SoejartoD D. (1985). Potential consequences of plant extinction in the United States on the current and future availability of prescription drugs. *Econ. Bot.*, 39(3):231-240.

Garg H S, (1993). Bioactive substance in marine algae, Marine biotechnology. Plenum Press, New York, pp. 1-8.

Goldfrank L , (1982). The Pernicious Panacea: Herbal Medicine. Hospital Physician 10, 64–86.

Haihua Shu, Hideko Arita, Masakazu Hayashida, Hiroshi Sekiyama and Kazuo Hanaoka, (2006). Effects of processed *Aconiti* tuber and its ingredient alkaloids on the development of antinociceptive tolerance to morphine. *J Ethanopharmacol.*, 103: 398–405.

Horborne J B, (1982). Introduction to Ecological biochemistry. 2nd edn Academic Press. London, New York, 278 pp.256-234.

Huss U, Ringbom T, Perera P, Bohlin L and Vasange M, (2002). Screening of ubiquitous plant constituents for COX-2 inhibition with a scintillation proximity based assay. *J Nat Prod.*, 65: 1517–1521.

Janssen P, Neimemegeers C J E and Dony J G H, (1963). The inhibitory effects of Fentanyl and other morphine like analgesics on the warm water induced tail withdrawal reflex in rats. *Arzneimittel Forschung*, 13: 502-507.

Jones J R, Ehrenfried L M, Saravanapavan P, (2006). Controlling ion release from bioactive glass foam scaffolds with antibacterial properties., *J Mater Sci Mater Med.*, 17: 989-996,

Kaboli P J, Doebbeling B N, Saag K G, Rosenthal G E, (2001). Use of complementary and alternative medicine by older patients with arthritis: a population-based study. *Arthritis Rheum.*, 45: 398-403.

Kim Y K, Kim R G, Park S J, Ha J H, Choi J W, Park H J and Lee K T, (2002). *In-vitro* anti-inflammatory activity of kalopanaxsaponin a isolated from *Kalopanax pictus* in murine macrophage RAW264.7 cells. *Biol. Pharm.* Bull., 25: 472–476.

Lanhers C M, Fleurentin J, Mortier F, Vinche A and Younos C, (1992). Anti-inflammatory and analgesic effects of an aqueous extract of *Harpagophytum procumbens, Plant Med,* 58: 117-123.

Lee D M, Weinblatt M E, (2001). Rheumatoid arthritis. *Lancet*, 358: 903- 911.

Levin D A and York B M,(1978). The toxicity of plant alkaloids: an ecogeographic perspective. *Biochem. System.Ecol.*, 6:61-76.

Li S H and Chu Y, (1999). Anti-inflammatory effect of total saponins of *Panax notoginseng*. *Zhongguo Yao Li Xue Bao,* 20: 551–554.

Mentz L A, Schenkel, E P, (1989). A coerenciae a confiabilidade das indicac¸o˜es terapeuticas. *Caderno de Farma´cia* 5 (1/2), 93–119.

Muruganandan S, Srivastava K, Chandra S, Tandan S K, Lal J and Raviprakash V., (2001). Anti-inflammatory activity of *Syzygium cumini* bark. *Fitoterapia,* 72:369–375.

Payne G, Bringi V, Prince C, Shuler M, (1991). Plant Cell and Tissue Culture in Liquid Systems. *Hanser Pub.* New York.

Ravikumar S, Abideen S, Syedali M K and Babuselvam M ,(2011). *In vitro* human sperm immobilizing activity of marine halophytes, *J. Phar. Res.,* 4(4): 1291-1293.

Ravikumar S, Anburajan S , Ramanathan G and Kaliaperumal N,(2002). Screening of seaweed extracts against antibiotic resistant post operative infection pathogens. *Seaweed Res. Util.,* 24:95-99.

Reinhold J G , Garcia S L and Garzan P, (1981). Binding of iron by fibre from wheat and maize. *Am. J. Clin. Nutr.*, 34: 1384-1391.

Singh G,(1998). Recent considerations in nonsteroidal anti-inflammatory drug gastropathy. *Am. J. Med.*, 105(1B): 315-385.

Soejarto D D, Farnsworth N R, (1989). Tropical rain forests: potential source of new drugs? *Perspect Biol Med.* Winter., 32(2):244–256.

Suh N, Honda T, Finlay H J, Barchowsky A, Williams C , Benoit N E, Xie J W, Nathan C , Gribble G W, and Sporn M B, (1998). Novel triterpenoids suppress inducible nitric oxide synthase (iNOS) and inducible cyclooxygenase (COX-2) in mouse macrophages. *Cancer Res.* 58: 717-723.

Swell RDE and Spencer P S J, (1976). Antinociceptive activity of narcotic agonist and partial agonist analgesics and other agents in the tail immersion test in mice and rats. *Neuropharmacol,* 15: 23-29.

Theophile D, Laure N E, Benoit N T, Anatole A G B, Emmanuel A A, Paul T V and Pierre K, (2006). Antinociceptive and anti-inflammatory effects of the ethyl acetate stem bark extract of *Bridelia scleroneura* (Euphorbiaceae). *Inflam Pharmacol.*, 14: 42–47.

Vazquez B, Avila G , Segura D and Escalante B, (1996). Anti-inflammatory activity of extracts from *Aloe vera* gel. *J Ethanapharmacol.*, 55:69-75.

Vorkoc J and Sedmera P, (1977). Evaluation of anti-inflammatory activity and toxicity related studies of *chloroxylon sweitenia. Phytochem.,* 11:2647.

Vulto A G, Smet, P A G M, (1988). In: Dukes, M.M.G. (Ed.). Meyler's Side Effects of Drugs, 11th Ed. Elsevier, Amsterdam, pp. 999–1005.

Winter C A and Poster C C, (1957). Effect of alteration inside chains upon anti-inflammatory and liver glycogen activities in hydrocortisone ester. *J Amer Pharmacol Soc,* 46:515–519.

Winter C A, Riseley E A, Nuss G W, (1962). Carrageenin induced edema in hind paw of the rat as an assay for anti-inflammatory drug. *Proceeding of Society Experimental Biology and Medicine,* 111: 544-547.

Wolf M, Lichtensteein D, Singh G, (1999). Gastrointestinal toxicity on nonsteroidal anti-inflammatory drugs. *New. Engl. J. Med.*, 340(24):1888-1889.

29

Phytochemical Screening from the Leaf Extracts of Medicinal Plants of Western Ghats, the Nilgiris

S. Padmavathy and V. Mekala

1. INTRODUCTION

Plant-derived substances have recently become of great interest owing to their versatile applications. Medicinal plants are the richest bio-resource of drugs of traditional systems of medicine, modern medicines, nutraceuticals, food supplements, folk medicines, pharmaceutical intermediates and chemical entities for synthetic drugs Medicinal value lies in certain chemical substances that have its physiological action on the human body. Many indigenous medicinal plants are used as spices and food plants. The general techniques of medicinal plant extraction include maceration, infusion, percolation, digestion, decoction, hot continuous extraction (Soxhlet), aqueous-alcoholic extraction by fermentation, counter-current extraction, microwave-assisted extraction, ultrasound extraction (sonication), supercritical fluid extraction, and phytonic extraction (with hydrofluorocarbon solvents). For aromatic plants, hydrodistillation techniques (water distillation, steam distillation, water and steam distillation), hydrolytic maceration followed by distillation, expression and enfleurage (cold fat extraction) may be employed.

Some of the latest extraction methods for aromatic plants include headspace trapping, solid phase micro- extraction, protoplast extraction, microdistillation, thermomicrodistillation and molecular distillation. The present study was carried out to identify the active chemical composition in the leaves of *Anaphalis beddomei, Piper wightii* and *Helichrysum hookerianum*

2. MATERIALS AND METHODS

Collection of plant materials

The plants were collected from the Western Ghats, The Nilgiris. The taxonomic identities of these plants were confirmed by Botanical Survey of India. Coimbatore. The plant leaves were air dried under shade, powered, sieved and the fine powder was used for extraction procedure and other evaluation.

Extraction procedure

Powdered leaves of the plants were soaked in six different solvents including Petroleum ether, Chloroform, ethyl acetate, Methanol, Water and Hexane in 1:2 ratio and kept for 2 days. The filtrate obtained was then used for phytochemical analysis.

Phytochemical screening

Each extract of *Anaphalis beddomei, Piper wightii* and *Helichrysum hookerianum* was subjected to preliminary phytochemical screening, to identify the chemical constituents of plant. The methods of analysis employed were standard procedures of Kokate.C.K and Harborne for the presence of various active components.

Detection of Alkaloid

Extracts were treated with Mayer's reagent and formation of yellow coloured precipitate indicates presence of alkaloids.

Detection of Saponin

Extracts were diluted with 20ml of distilled water and shaken in a graduated cylinder for 15 minutes. Formation of 1 cm layer of foam indicates the presence of saponins.

Detection of Tannins

To 2ml of the extract, few drops of 1% lead acetate was added and formation of yellowish precipitate indicates presence of tannins.

Detection of Steroid

1ml of extract with acetic acid, chloroform and sulphuric acid resuls in violet to blue green formation.

Detection of Flavanoid

Extract with Sulphuric acid results in orange colour.

Detection of Terpenoid

2ml of extract with 2ml of acetic acid and sulphuric acid results in blue green rings.

Detection of Phloba tannins

Extract boiled with 1% hydrochloric acid results in red precipitate deposition.

Detection of Coumarin

2ml of extract with 10% of 3ml sodium hydroxide results in yellow colour.

Detection of Cycloglycosides

5ml of extract 2ml of acetic acid ,1% ferric chloride 1 drop and sulphutic acid 1ml turns to brown violet and greenish ring formation.

Detection of Total phenol

Extract with 3% ferric chloride results in deep blue.

Detection of Quinone

1gm of extract with 5ml of hydrochloric acid results in yellow precipitate.

3. RESULTS AND DISCUSSION

The phytochemical analysis of the three endangered plants investigated is tabulated. Alkaloid, saponin, tannin, steroids, flavonoid and terpenoid are present in methanolic extract of all the three plants.

The leaf extracts of *A beddomei* when treated with Petroleum ether revealed the presence of steroids and terpenoids. The presence of steroid was also confirmed by Chloroform extract. Further, chloroform extract indicated the presence of saponins and flavanoids in the leaf extracts of Anaphalis. The presence of saponins indicates, antidiahorreal, antihelminthic and anticancer activity of the plant. Ethyl acetate extract indicated the presence of tannins in addition to steroids, flavanoids and terpenoids. Except Phlobatannins, cardioglycosides and total phenolic content the leaf extracts of *A beddomei* indicated the presence of all other phytochemical under study. More specifically, the presence of alkaloids and tannins indicates antimicrobial, antidiahorreal and antihelminthic properties in the plants of *A beddomei*. Also, the presence of Quinones indicated the presence of anti-microbial activity. The water extract indicated the presence of tannins, steroids, terpenoids, coumarins, cardioglycosides and total phenol content in the plant. Finally, extraction with hexane confirmed the presence of steroids and terpenoids in the plant.

Table 1: Preliminary phytochemical screening of Leaf extracts of *Anaphalis beddomei*

	Alkaloid	Saponin	Tannin	Steroid	Flavonoid	Terpenoid	Pholo	Coumarin	Cardio	Total phe	Quinone
Petroleum ether	-	-	-	+	-	+	-	-	-	-	-
Chloroform	-	+	-	+	+	-	-	-	-	-	-
Ethyl acetate	-	-	+	+	+	+	-	-	-	-	-
Methanol	+	+	+	+	+	+	-	+	-	-	+
Water	-	-	+	+	-	+	-	+	+	+	-
Hexane	-	-	-	+	-	+	-	-	-	-	-

Table 2: Preliminary phytochemical screening of extracts of *Piper wightii*

	Alkaloid	Saponin	Tannin	Steroid	Flavonoid	Terpenoid	Pholo	Coumarin	Cardio	Total phe	Quinone
Petroleum ether	-	-	-	+	+	-	-	-	-	-	-
Chloroform	-	-	-	+	+	+	-	-	-	-	-
Ethyl acetate	-	-	-	+	+	+	-	-	-	-	-
Methanol	Š	Š	+	+	+	+	-	+	-	-	+
Water	-	-	+	-	-	+	-	+	+	+	-
Hexane	-	-	-	+	+	+	-	-	-	-	-

Table 3: Preliminary phytochemical screening of Leaf extracts of *Helichrysum hookerianum*

	Alkaloid	Saponin	Tannin	Steroid	Flavonoid	Terpenoid	Pholo	Coumarin	Cardio	Total phe	Quinone
Petroleum ether	-	-	-	+	-	+	-	-	-	-	-
Chloroform	-	+	-	+	+	-	-	-	-	-	-
Ethyl acetate	-	-	-	-	+	+	-	-	-	-	-
Methanol	+	+	+	+	+	+	-	+	-	+	+
Water	-	-	+	+	+	+	-	+	+	+	-
Hexane	-	-	-	+	+	+	-	-	-	-	-

The leaf extracts of *P wightii* when treated with Petroleum ether revealed the presence of steroids and flavanoids. The presence of flavanoids indicated antimicrobial and antidiarrhoeal activity of *P wightii*. The presence of steroid and flavonoids was also confirmed by hexane, chloroform and ethyl acetate extracts. Further, Hexane, chloroform and ethyl acetate extracts indicated the presence of terpenoids in the leaf extracts of *P wightii*. Except Phlobatannins, cardioglycosides and total phenolic content the leaf extracts of *P wightii* indicated the presence of all other phytochemical under study. Finally, the water extract confirmed the presence of tannins, terpenoids, coumarins, cardioglycosides and total phenols in the plant.

The leaf extracts of *H hookerianum* when treated with petroleum ether revealed the presence of steroids and terpenoids. The presence of steroid was also confirmed by Chloroform extract. Further, chloroform extract indicated the presence of Saponins and Flavanoids in the leaf extracts of the plant. The presence of saponins indicates, antidiahorreal, antihelminthic and anticancer activity of the plant. Ethyl acetate extract indicated the presence of terpenoids in addition to flavanoids. Except Phlobatannins, cardioglycosides in the leaf extracts of *H hookerianum* indicated the presence of all other phytochemical under study. More specifically, the presence of alkaloids, tannins and total phenol indicates antimicrobial, antidiahorreal and antihelminthic property in the plants of *H hookerianum*. The water extract indicated the presence of tannins, steroids, flavonoids, terpenoids, coumarins, cardioglycosides and total phenol content in the plant. Finally, extraction with hexane confirmed the presence of steroids, flavonoids and terpenoids in the plant.

4. CONCLUSION

It is found from the present study that *A beddomei, P wightii* and *H hookerianum* contain many phytochemicals like alkaloids, flavonoids, steroids, saponins, coumarins, phenols, terpenoids, and quinones. These results are suggestive of primary bioactive compound of commercially importance and may result in great interest in plants pharmaceuticals. Primary metabolites analysis is necessary for knowing the nutritional potential of plants and them also from the precursors for the synthesis of secondary metabolites.

REFERENCES

Cowan MM. Plant products as antimicrobial agents. *Clinical Microbiology Reviews* 1999; 12(4): 564- 582.

Das K, Tiwari RKS, Shrivastava DK. 2010. Techniques for evaluation of medicinal plant products as antimicrobial agent: Current methods and future trends. *Journal of Medicinal Plants Research*, 4(2).

Handa SS, Khanuja SPS, Longo G, Rakesh DD. 2008. Extraction Technologies for Medicinal and Aromatic Plants. International Centre for Science and High Technology, *Trieste*, 21-25.

Harbone J.B. 1999. Phytochemical Methods, Chapman & Hall, London, pp.60 66.

Kokate C.K, 2000. Practical Pharmacognosy, Vallabh Prakashan, Delhi, pp.107 111.

Kumar R, Sharma RJ, Bairwa K, Roy RK, Kumar A. 2010. Pharmacological review on natural antidiarrhoel agents. *Der Pharma Chemica*, 2(2): 66-93.

Ncube NS, Afolayan AJ, Okoh AI. 2008. Assessment techniques of antimicrobial properties of natural compounds of plant origin: current methods and future trends. *African Journal of Biotechnology*, 7(12):1797-1806.

Okwu, DE. 2001. Evaluation of the chemical composition of indigenous spices and flavouring agents. Global *J. Pure Applied Sci.* 7:455-549.

Sutar N, Garai R, Sharma US, Sharma UK. 2010. Anthelmintic activity of Platycladus orientalis leavesextract. *International Journal of Parasitology Research,* 2(2): 1-3.

Wang GS, Han J, Zhao LW, Jiang DX, Liu YT, Liu XL. 2010. Anthelmintic activity of steroidal saponins from Paris polyphylla. *Phytomedicine*, 17: 1102-1105.

30

Screening of Common Leafy Vegetables of Tamil Nadu for their Nutraceutical and Antioxidant Properties

Shanmugaraj Bala Murugan, Ramamoorthy Deepika, Aziz Reshma and Ramalingam Sathishkumar

1. INTRODUCTION

Phytochemicals are biologically active compounds found in plants, which act as a resistance system for host plants and protect them from biotic and abiotic stresses. Phytochemicals act as antioxidants, which react with the free radicals present in the cell. Free radicals are formed in the cell as a result of aerobic metabolism and these radicals are biologically active that damages the macromolecules like DNA, proteins and lipids resulting in oxidizing chain reactions (Moussa *et al.*, 2011). Oxidative stress damage leads to various disorders like autoimmune disorders, cancer, rheumatoid arthritis, cataract, aging, cardiovascular diseases and other neurodegenerative disorders (Uttara *et al.*, 2009). The major actions of antioxidant in the cells are to protect them against the damage caused by reactive oxygen species (ROS).

Medicinal plants are good sources of antioxidants such as vitamin E, ascorbic acid and carotenoids. There are huge numbers of medicinal plants that are veiled in the environment. In ancient times, different plant species have been used for the treatment of human ailments from fever to cancer (Devendrakumar *et al.*, 2009). Since the evolution of human race, the diseases also have start-making headway hence traditional people use locally available plants to cure various disorders. Even though humans can develop modern health care systems/synthetic drugs, still humans are a step backward to nature; unpredictable diseases are also evolved in our day-to-day life that will challenge the humanity. Hence, the humans are in search of effective drugs that will combat deadly diseases without any side effects and with low cost. Modern medical care is not affordable by the rural mass due to the cost factor. Earlier food has been viewed as a source of medicine as well as nutrition. Synthetic antioxidants like butylated hydroxyanisole (BHA) and butylated hydroxytoluene (BHT) have more side effects hence they are banned. Screening of medicinal herbs has now become a probable basis for the isolation of bioactive compounds and natural antioxidants since the medicinal plants are the reservoir of many bioactive compounds that has potential health and pharmaceutical applications. Hence, interest has been shifted towards the documentation of these plants that has potential health benefits (Singh *et al.*, 2011).

Now it is essential to renew the importance of some neglected green leafy vegetables that have potential medicinal properties. In ancient system of medicine, *Trigonella foenum-graecum* is used to cure stomach pain, nervous disorders, dyspepsia and ulcer. The stem of *Mentha arvensis* is used in tooth pastes as a flavouring agent in pharmaceuticals and also used for the treatment of stomach disorders like indigestion, acidity. The leaves and stem of *Centella asiatica* are widely used for the treatment of skin diseases, leprosy, nerves and blood; leaves useful for improving memory and in syphilitic skin diseases (Panda and Misra, 2011). Leaves of *Murraya koenigii* can be used for the treatment of dyspepsia and dysentery (Buragohain, 2011). Due to these, the green leafy vegetables are regarded as 'Natures anti-aging wonders' (Gupta and Prakash, 2009). Thus the aim of this study is to revitalize the nutritional attributes of green leafy vegetables as it is known to have several antioxidant properties.

2. MATERIALS AND METHODS

Vegetable samples

Eight leafy green vegetable samples namely *Amaranthus paniculatus, Amaranthus polygonoides, Centella asiatica, Mentha arvensis, Murraya koenigii, Solanum nigrum, Spinacia oleracea* and *Trigonella foenum – graecum* were collected from the local market of Coimbatore, Tamil Nadu, India.

Sample extraction

The vegetables were cleaned and cut into small pieces. 0.5 g of sample was weighed and homogenized in 5 ml each of methanol, until the tissue gets into a fine paste in a mortar and pestle. The samples were then centrifuged at 4000 rpm for 10 min. The supernatant was collected and stored at -20°C for further analysis.

Phytochemicals

Estimation of total flavonoid content

Total flavonoids were quantified by aluminium chloride method (Chang *et al.*, 2002). To 1 ml of methanolic extract, 0.5 ml of 2% w/v $AlCl_3$ in methanol and 0.5 ml potassium acetate (120mM) were added and incubated at room temperature for 30 minutes. Absorbance was read at 415 nm. Quercetin was used as a standard and the results were expressed as mg of quercetin equivalents per 100 g of fresh weight sample.

Determination of total phenolics

Methanolic extract (100 µL) was mixed with 1 ml of 10% Folin-Ciocalteau reagent and it was incubated at room temperature for 4 minutes. Then 2 ml of 5% Sodium carbonate was added to above mixture and vortexed. The resultant mixture was incubated in dark for 45 minutes at room temperature. Following this, the absorbance of the sample was measured at 765 nm using gallic acid (100-1000 µg/ml) as a standard. Results were expressed as mg of gallic acid equivalents per 100 g of fresh weight of sample (Singleton and Rossi, 1965).

Antioxidant capacity

DPPH radical scavenging activity

The hydrogen donating ability of the plant extract was evaluated using stable free radical DPPH according to the method of Gyamfi *et al.*, (1999). Methanolic extract (100 µL) of sample was mixed with 900 µL of Tris HCl buffer (50mM, pH = 7.4) and 2 ml of DPPH (0.1mM in methanol). DPPH working solution was prepared freshly. The solution was incubated at room temperature for 30 minutes and the absorbance was read at 517 nm. The percentage of DPPH scavenging activity was determined using the formula,

DPPH Radical Scavenging Activity (%) = $[(A_0 - A_1)/A_0]$

Where A_0 is the absorbance of control and A_1 is the absorbance of sample.

Ferric Reducing Antioxidant Power Assay

The ferric reducing antioxidant activity was determined by a modified method of Benzie and Strain (1996). The FRAP reagent was prepared by mixing 25 ml of 300mM Acetate buffer (pH = 3.6), 2.5 ml of 10mM TPTZ solution in 40mM HCl and 2.5 ml of 20mM $FeCl_3.6H_2O$ solution. The solution was pre-warmed to 37°C before use. Methanolic extract (100 µL) were allowed to react with 2 ml of the FRAP solution, and the change in the absorbance was calculated after 30 minutes at 593 nm using $FeSO_4$ (200- 1000µM) as reference standard.

Statistical analysis

Each assay was repeated 3 times for all the samples and the data represented are the means of 3 replicates. Analysis of Variance (ANOVA) followed by Duncan's multiple range tests was used to test the significant differences among the samples ($p<0.05$).

3. RESULTS AND DISCUSSION

One of the major setbacks for the evaluation of antioxidant property of the plant extract is the choice of standards and analysis methods, because all the methods are based on the inhibition of substrate oxidation initiated by free radicals as in the case of hydroxyl, superoxide and peroxyl radicals or based on the scavenging ability of synthetic, stable free radicals like DPPH and ABTS. Hence, two or more methods are always used to assess the antioxidant ability of the plant extract for the purpose of screening (Hassimotto *et al.*, 2005). In this study, phytochemical content can be assessed by aluminium chloride and Folin's Ciocalteau method; the radical scavenging ability and reducing power of the extract have been evaluated by DPPH and FRAP method respectively.

Phenolic acids and flavonoids are the major class of secondary metabolites that possess antioxidant activity. In the present study, the amount of total flavonoids in the leaf fraction of the green leafy vegetables ranged from 59.05-75.32 mg of quercetin equiv/100 g of fresh weight (Fig. 1a). *Murraya koenigii* had the highest total flavonoid content among the 8 leafy vegetables studied. Phenolic acids are the major determinants of antioxidant properties of vegetables due to its redox properties, ability to quench the singlet free radicals etc. Phenolic contents of 8 selected leafy vegetables are expressed as mg of gallic acid equiv/100 g of fresh weight in Fig. 1b. The total phenolic content of the leafy vegetable investigated varied from 110.76-374.82 mg of Gallic acid equiv/100 g of fresh weight. Among all the vegetables analyzed, *Murraya koenigii* had the highest, while *Amaranthus paniculatus* has the lowest phenolic content. Gupta and Prakash (2009) reported the total phenolic content of *Murraya koenigii* and *Trigonella foenum graecum* to be 387.5 and 158.3 mg of tannic acid equiv/100 g of FW, whereas Kaur and Kapoor (2002) showed that the total phenolic content of *Trigonella foenum graecum* to be 217.5 mg of catechol equiv/100 g of FW. The difference in the flavonoids and phenolic content of the vegetables may be due to the biotic and abiotic stress besides difference in the genetic, agronomic and environmental factors like temperature, climate, storage conditions and post harvest handling (Howard *et al.*, 2003; Islam *et al.*, 2003; Young *et al.*, 2005; Zhao *et al.*, 2009).

DPPH and FRAP method are the oldest and most frequently used methods for assessing the total antioxidant capacity of the vegetables. DPPH is the stable, synthetic nitrogen radical used for assessing the hydrogen donating ability of the phenolic compounds. It is a purple colored compound, by the addition of the extracts having the hydroxyl groups will reduce/scavenge the DPPH, hence the color gradually changes to green/yellow and the decrease in the absorbance was monitored at 517 nm (Subashree *et al.*, 2009). The results of DPPH radical scavenging activity of the vegetable extracts are given in Figure 1c. The highest DPPH radical scavenging activity was found in the *Murraya koenigii* (89.43%) while the lowest was in *Centella asiatica* (49.78%). Our results are in agreement with Gupta and Prakash (2009), where they have showed *M. koenigii* had the strongest radical scavenging activity (83.44%) among the four vegetables studied.

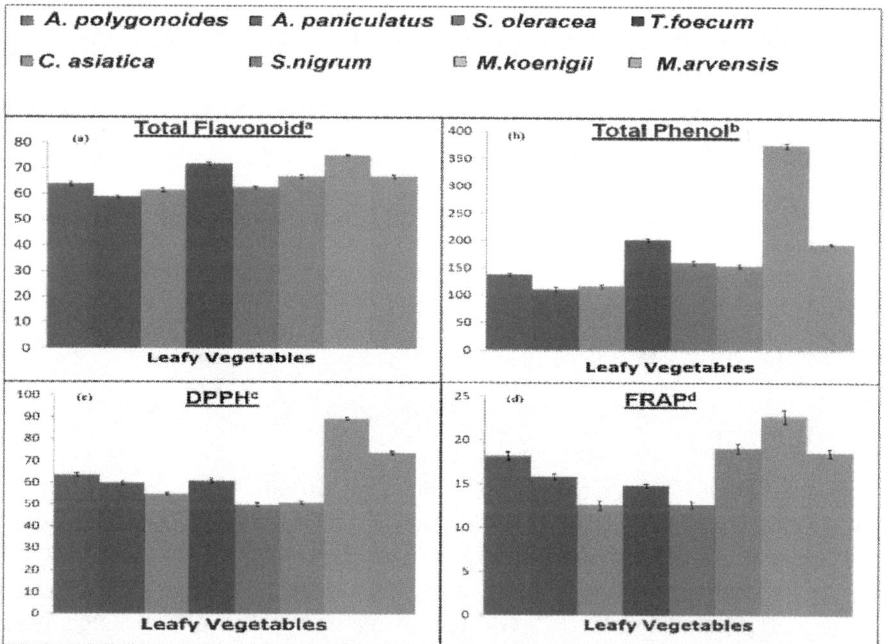

[a]Data expressed as milligrams of Quercetin equivalents per 100 g fresh weight.
[b]Data expressed as milligrams of Gallic acid equivalents (GAE) per 100 g fresh weight .
[c]Data expressed as percentage of DPPH radical scavenging activity.
[d]Data expressed as mM of ferrous equivalent Fe (II) per 100 g fresh weight.

Fig.1: Phytochemical and radical scavenging activities of 8 green leafy vegetables

The reducing power of the vegetable extracts was expressed in terms of mM of Ferrous Fe (II) equiv/100g fresh weight of the sample (Figure 1d). The results obtained in the present study showed that the *Murraya koenigii* (22.75 mM) had highest FRAP values; while all the other samples have significant FRAP values ranging from 12.56 to 22.75 mM. Similar observations were reported by Katerere *et al.*, (2012) on African green leafy vegetables.

According to the results obtained (Table 1), strong correlation was found between the DPPH radical scavenging activity and phenolic content (R=0.836) of the vegetables, whereas no significant correlation was found between the FRAP values with the phytochemical content. Our data indicated that the radical scavenging activity of the vegetable extract was determined by their phenolic content. Several authors found the strong correlation between phenolic acids and antioxidant activity in various food commodities (Eberhardt *et al.*, 2000; Cai *et al.*, 2004; Katerere *et al.*, 2012), while others found no such correlation (Kahkonen *et al.*, 1999; Dasgupta and De, 2007) because the antioxidant activity of phenolic acids are determined by its structure, position and availability of hydroxyl groups (Stratil *et al.*, 2006). These results are in agreement with observations from other studies and support the fact that antioxidant activity of the vegetable is not due to any compound; it may be due to the synergistic and antagonistic effects of all the secondary metabolites.

Table 1: Correlation coefficient between total Flavonoid, Phenolics and radical scavenging activities of green leafy vegetables (n = 8 samples)

R^a	Flavonoid	Phenolics
DPPH	0.665^{NS}	0.836**
FRAP	0.637^{NS}	0.694^{NS}

^{a}R, Correlation Coefficient.
**Significant level at P < 0.01; NS-Not Significant

4. CONCLUSION

From these results, we can conclude that the common culinary vegetable *Murraya koenigii* showed the highest antioxidant activity. It has been found that the green leafy vegetables have considerable antioxidant properties that can be help us as a dietary antioxidant supplement and also helps to manage the age related human ailments. This finding also helps to recuperate our ancestral knowledge about these underutilized vegetables, which can be used to develop functional food through genetic engineering approaches in order to meet the future nutrition and health challenges. This study could be the prelude for the concept "Let food be your medicine and medicine be your food".

REFERENCES

Benzie, I.F., Strain, J.J., 1996. The ferric reducing ability of plasma (FRAP) as a measure of "antioxidant power": *The FRAP Assay. Analytical Biochemistry* 239, 70-76.

Buragohain, J., 2011. Ethnomedicinal Plants Used by the ethnic Communities of Tinsukia District of Assam, India. *Recent Research in Science and Technology* 3, 31-42

Cai, Y., Luo, Q., Sun, M., Corke, H., 2004. Antioxidant activity and phenolic compounds of 112 traditional Chinese medicinal plants associated with anticancer. *Life Sciences* 74, 2157-2184

Chang, C.C., Yang, M.H., Wen, H.M., Chern, J.C., 2002. Estimation of total flavonoid content in Propolis by two complementary colorimetric methods. *Journal of Food and Drug Analysis* 10, 178-182.

Dasgupta, N., De, B., 2007. Antioxidant activity of some leafy vegetables of India: A comparative study. *Food Chemistry* 101, 471-474.

Devendrakumar, D., Anbazhagan, M., Gomathi, M., Rajendran, R., 2009. Traditional Phytotherapy of diabetes used by the people of Perumbular district, Tamilnadu, South India. *Recent Research in Science and Technology* 1, 287-290.

Eberhardt, M.V., Lee, C.Y., Liu, R.H. 2000. Antioxidant activity of fresh apples. *Nature* 405, 903"904.

Gupta, S., Prakash, J., 2009. Studies on Indian green leafy vegetables for their antioxidant activity. *Plant Foods for Human Nutrition* 64, 39–45.

Gyamfi, M.A., Yonamine, M., Aniya, Y., 1999. Free radical scavenging action of medicinal herbs from Ghana: Thonningia sanguine on experimentally-induced liver injuries. *General Pharmacology* 32, 661-667.

Hassimotto, N.M.A., Genovese, M.I., Lajolo, F.M., 2005. Antioxidant activity of dietary fruits, vegetables and commercial frozen fruit pulps. *Journal of Agricultural and Food Chemistry* 53, 2928–2935.

Howard, L.R., Clark, J.R., Brownmiller, C., 2003. Antioxidant capacity and phenolic content in blueberries as aûected by genotype and growing season. *Journal of the Science of Food and Agriculture* 83, 1238–1247

Islam, M.S., Yoshimoto, M., Ishigure, K., Okuno, S., Yamakawa, O., 2003. Eûect of artiûcial shading and temperature on radical scavenging activity and polyphenolic composition in sweetpotato (*Ipomoea batatas* L.) leaves.*Journal of the American Society of Horticultural Science* 128, 182–187.

Kahkonen, M.P., Hopia, A.I., Vuorela, H.J., Rauha, J.P., Pihlaja, K., Kujala, T.S., Heinonen, M., 1999. Antioxidant activity of plant extracts containing phenolic compounds. *Journal of Agricultural and Food Chemistry* 47, 3954"3962

Katerere, D.R., Graziani, G., Thembo, K.M., Nyazema, N.Z., Ritieni, A., 2012. Antioxidant activity of some African medicinal and dietary leafy African vegetables. *African Journal of Biotechnology* 11, 4103- 4108.

Kaur, C., Kapoor, H.C., 2002. Anti-oxidant activity and total phenolic content of some Asian vegetables.*International Journal of Food Science and Technology* 37, 153–161.

Moussa, A.M., Emam, A.M., Diab, Y.M., Mahmoud, M.E., Mahmoud, A.S., 2011. Evaluation of antioxidant potential of 124 Egyptian plants with emphasis on the action of *Punica granatum* leaf extract on rats. *International Food Research Journal* 18, 535–542.

Panda, A., Misra, M.K., 2011.Ethanomedicinal survey of some wetland plants of south Orissa and their conservation. *Indian Journal of Traditional Knowledge* 10, 296-303

Singh, A.G., Poudel K.N., Tewari, D.D., 2011. Diversity of Cultivated and Wild Medicinal Plants Used by People of Devdaha VDC of Rupandehi District, West Nepal.*Current Botany* 2, 34-42.

Singleton, V.L., Rossi, J.A., 1965. Colorimetry of total phenolics with phosphomolybdic-phosphotungstic acid reagents. *American Journal of Enology and Viticulture* 16, 144-158.

Stratil, P., Klejdus, B., Kuban, V., 2006.Determination of total content of phenolic compounds and their antioxidant activity in vegetables—evaluation of spectrophotometric methods.*Journal of Agricultural and Food Chemistry* 54, 607-616.

Subhasree, B., Baskar, R., Keerthana, R.L., Susan, R.L., Rajasekaran, P., 2009. Evaluation of antioxidant potential in selected green leafy vegetables. *Food Chemistry* 115, 1213–1220.

Uttara, B., Singh, A.V., Zamboni, P., Mahajan, R.T., 2009. Oxidative stress and neurodegenerative diseases: a review of upstream and downstream antioxidant therapeutic options. *Current Neuropharmacology* 7, 65-74.

Young, J.E., Zhao, X., Carey, E.E., Welti, R., Yang, S.S., Wang, W.Q., 2005. *Phytochemical* phenolics in organically grown vegetables. *Mol. Nutr. Food Res* 49, 1136-1142.

Zhao, X., Nechols, J.R., Williams, K.A., Wang, W.Q., Carey, E.E., 2009. Comparison of phenolic acids in organically and conventionally grown pakchoi (*Brassica rapa* L. *chinensis*). *J. Sci. Food Agric* 89, 940-946.

MIX
Papier | Fördert gute Waldnutzung
FSC® C083411

Zeitfracht Medien GmbH
Ferdinand-Jühlke-Straße 7
99095 Erfurt, Deutschland
produktsicherheit@kolibri360.de